800 MILES TO VALDEZ

James P. Roscow

800 MILES

TO VALDEZ

The Building of the Alaska Pipeline

PRENTICE-HALL, INC., Englewood Cliffs, N.J.

(**Front Endpaper**) Permafrost—soil and other ground material that remains frozen for two years or longer—was found along 85% of the pipeline route.

(**Back Endpaper**) The first section of the pipeline was buried deep under the Tonsina River in southern Alaska in March of 1975. The pipeline passes over more than 800 streams and rivers along its route, requiring some 30 miles of buried and elevated river crossings.

Portions of this book originally appeared, in slightly different form, in *Banking, Energy, Exchange, Financial World, Signature,* and *Alyeska Reports* magazines.

800 Miles to Valdez: The Building of the Alaska Pipeline, by James P. Roscow
Copyright © 1977 by James P. Roscow
All rights reserved. No part of this book may be reproduced in any form or by any means, except for the inclusion of brief quotations in a review, without permission in writing from the publisher.
Printed in the United States of America
Prentice-Hall International, Inc., London
Prentice-Hall of Australia, Pty. Ltd., Sydney
Prentice-Hall of Canada, Ltd., Toronto
Prentice-Hall of India Private Ltd., New Delhi
Prentice-Hall of Japan, Inc., Tokyo
Prentice-Hall of Southeast Asia Pte. Ltd., Singapore
Whitehall Books Limited, Wellington, New Zealand
10 9 8 7 6 5 4 3 2 1

Library of Congress Cataloging in Publication Data

Roscow, James P
  800 miles to Valdez.
  Includes index.
  1. Alaska pipeline—History. I. Title
TN879.5.R55   338.4'7'66554   77-24363
ISBN 0-13-246835-2

for Nina, Stephen, and April . . .
who ask good questions

Contents

800 MILES TO VALDEZ

Prologue:
All of the Issues
of Our Time

"You're going to Alaska to write about the pipeline? I've been read-ing a lot about that in the papers," says the woman on the Carey bus to New York's John F. Kennedy International Airport one January morning. "Alaska must be an interesting place . . . I'd like to go there sometime."

Wistfully, she leaves the bus at the El Al terminal, bound only for her waiting desk. Her travels are done, at least for this day. Yours are not. Some eight hours and 4,600 miles later, Pan Ameri-can's 747 sinks like a feathery stone into the brisk 20°-below-zero early afternoon dusk of Fairbanks—and you start out to see what the Alaska pipeline is all about.

As you begin this book, a long-awaited, much-delayed, and still very controversial event is happening in a remote, beautiful land: Alaska. Oil—hundreds of thousands of barrels a day of raw, dark, fragrant crude oil—is making its way south inside 800 miles of 48-inch steel pipe from Prudhoe Bay on the edge of the Arctic Ocean to tankers anchored at Valdez in Prince William Sound just off the Gulf of Alaska. Fresh from the earth at a production tem-perature of about 160°, oil from Alaska's North Slope at last is coming to market.

1

Why should you care?

There are many reasons. Which of them may interest you enough to read a book about the Alaska pipeline will depend on a number of things about you. Whether you're an American, or an Arab—or an Alaskan. Whether you drive a car, heat a home, or work in a business that uses fuels. Whether you're interested in energy and other resources. Whether you're interested in the geopolitics of world oil—the interplay of international economies and governments. Whether you're interested in minority rights and the Indian movement, in the United States and perhaps in Canada. Whether you're interested in the confrontations between states' rights and federal policy. Whether you're a corporate manager or a member of a labor union. Whether you're a geologist, a biologist—or a conservationist. Whether you work for an oil company, a pipeline company, or a construction company—and which one.

Perhaps you like to hike, hunt, fish, or climb mountains. Perhaps you enjoy reading about big, complicated engineering projects. Perhaps you like science and technology, and how difficult technical problems are solved. Perhaps you were stationed in Alaska during World War II or afterward. Perhaps you once lived on a frontier. Perhaps you live on one now, or would like to. Or perhaps you're a travel buff who likes to go to colorful, far-off places—really go there, or go there in the pages of a book.

If you're one, or several, of these people, the new oil pipeline down through Alaska has meanings for you.

The oil field at Prudhoe Bay was discovered in early 1968. It is still the largest petroleum deposit in the United States. It's taken nearly ten years to start bringing this oil to market. It might have taken only four years. The Alaska pipeline has cost nearly $8 billion—and almost twice that much if you add in the financing charges and the expenses of developing the oil field. The cost of the pipeline alone might have been only a little more than half that. The pipeline is the largest and most expensive project ever undertaken by private industry. By the time it began operating in the summer of 1977, some 70,000 people from all over the world had worked on it. Technically, managerially, and legislatively, the pipeline has dwarfed any other modern-day industrial enterprise.

It has all but dwarfed the industry that built it. Oil companies, by the nature of their business, are among the largest companies in

All of the Issues of Our Time

the world. Yet the Alaska pipeline's total costs are larger than the assets of all but three of its eight owner companies. Plans for the pipeline were begun in 1968 by Atlantic Richfield Company and Humble Oil & Refining Company—now Exxon Company USA, the U.S. affiliate of Exxon Corporation. Later that year, the two partners joined with the other North Slope pioneer, British Petroleum Ltd., to form TAPS—the Trans Alaska Pipeline System. In 1969 five more companies joined TAPS: Mobil Oil Corporation, Phillips Petroleum Company, Union Oil Company of California, Amerada Hess Corporation, and Home Oil Company of Canada, which later dropped out of the project.

In 1970 the TAPS joint venture was succeeded by Alyeska Pipeline Service Company, a nonprofit corporation formed to design, build, and operate the pipeline system. Alyeska took for its name the ancient Aleut word for Alaska itself—to the natives, it meant "the Great Land." The seven remaining TAPS partners held undivided interests in the project, either in their corporate names or through their pipeline affiliates. Also in 1970 British Petroleum traded its North Slope leases to The Standard Oil Company (Ohio) for 25 percent of SOHIO's stock. SOHIO now became the eighth company in Alyeska, with the largest share of Prudhoe Bay's oil and the pipeline that would carry it.

Four of the Alyeska owner companies—SOHIO, Atlantic Richfield (ARCO), Exxon USA, and British Petroleum—own more than 90 percent of the pipeline and its oil. And each of them has been dramatically reshaped by its adventure in Alaskan oil. ARCO and SOHIO have been catapulted from the middle ranks of U.S. oil companies to among the biggest and best-balanced companies in the industry. British Petroleum, which was the seventh largest in worldwide oil and heavily dependent on the Middle East for its supplies, has now combined its successes in Alaska and the North Sea to become one of the three largest companies in international petroleum. And Exxon USA, even though it is part of the largest industrial company in the world, may soon gain nearly half of all of its U.S. oil from Alaska alone.

Superlatives, perhaps. But Alaska breeds superlatives. And so has the Alaska pipeline.

In cutting across some 800 miles of Alaskan terrain, much of it a timeless wilderness still known best to the state's native people, the pipeline also cuts across virtually every issue important to our society in the 1970s. Alaskan oil and the pipeline project have had a large part in defining several of these issues. They have even had a part—wittingly at times, unwittingly at other times—in resolving some of these issues. Some examples? To list them is to catalog what matters most to us today, a litany of our troubled and fascinating age.

The "energy crisis." The Arab oil embargo of late 1973 and the quintupling of the world price of oil since then have confirmed years of warnings about looming United States and worldwide energy problems. Whether we are dealing with a real "crisis," or some more complex situation that ought to have a quieter name, can still be argued. Whatever has been happening, it helped get the Alaska pipeline project moving again. It had been stopped dead, by 1973, for nearly three years. Now the pipeline has been built.

Worldwide energy balances. For better or worse, oil in the last century has been one of the great shaping forces in world history. Whole libraries exist of books on the economics and politics of world oil, and on the strategic implications of where oil is and who controls it. Ten years ago, Alaska's oil and natural gas were not at all an important part of the world energy picture. Today, they are.

U.S. energy supplies. Through World War II, the United States controlled more of the world's oil, and produced more of it, than any other nation or group of nations. Today, thirty years later, the U.S. is a poor fifth among world oil powers. The decline in its rate of production and world ranking is, if anything, still accelerating. Before 1968, Alaska's oil and gas were a sliver on any chart of U.S. supplies. From now on, Alaska may contribute the largest part of all U.S. reserves.

U.S. energy independence. To whatever extent this comes true, Alaskan energy—and its availability in major U.S. marketplaces— could stand behind any real hope the U.S. has of greater independence or self-sufficiency in energy supplies.

The environment. Alaska and the pipeline have been, without question, a main rallying point of today's still growing concern over ecology. Even as the pipeline was announced in 1969, Congress was preparing to pass the National Environmental Policy Act. As a result of NEPA, the pipeline has been built with environmental constraints and scrutiny—and high costs—never before imposed on an industrial project of any kind. Nor is it likely that any large project of the future, by industry or government, will be built any other way.

Arctic construction. At the same time, the building of the pipeline has proven to environmentalists themselves that the Arctic is not all that fragile. Nature in Alaska has been a brawny adversary for the construction crews. It is as resilient there as anywhere else against man's intrusion as well as its own upheavals. Alaska's High North all the way to the Arctic Ocean had been conquered by big military construction projects in earlier years. Much of that experience, mistakes and all, went into the early research on the Alaska pipeline. Those contractors would have liked to know what the pipeline builders now know of coping with the formidable construction conditions in lands around the Arctic Circle.

Pipeline technology. Big, challenging projects push forward their own technical horizons. There are fascinating analogies in the Alaska pipeline project and some massive efforts of the past: the first United States transcontinental railroads, the 2,500-mile Canadian Pacific Railway soon afterward, the Panama Canal early in this century, the Al-Can Highway in World War II—and, less than twenty years ago, the United States space program to put men on the moon. If the Alaska pipeline has advanced Arctic construction, even more so it has added to the future technology of building and operating pipelines everywhere.

Other resource development. None of these issues is only about oil. Substitute "natural resource" for "energy" and you have the same issues. They are worldwide, since resources are world commodities. They are United States issues, since the U.S. ranks large in so many resources. And they are surely Alaskan issues. Alaska's natural wealth is scarcely limited to its oil and gas. It has extensive timber in its southeastern panhandle, and a fishing industry on a grand scale in its far-reaching coastal waters. And Alaska is rich in undeveloped coal, copper, iron, nickel, and yet more of the gold that touched off its first resource boom eighty years ago. So Alaska is a prime focus of today's worldwide resource debate—whether to learn to live with less or to range out to the far frontiers to develop everything that is left.

Nor are the issues that touch on Alaska and the pipeline only about resources. They extend as well into social, economic, and political matters.

Native rights. Who would have thought that an industrial scheme like the pipeline would lead directly to resolving a major minority rights issue? Yet it did. Alaskan native land claims had stood for centuries, since Russia, Europe, and the United States began to exploit the new land in the northwest Pacific. The push to develop North Slope oil coincided exactly with the final push to settle the land claims. The chance to leverage social needs against business ambitions was obvious and irresistible. The result was the Alaska Native Claims Settlement Act of 1971. ANCSA restored 44 million acres of land to Alaska's 70,000 natives—Aleuts, Indians, and Eskimos. It gave them long-term investment cash, much of it in the form of a stake in Alaska's royalties from North Slope production. And Alaskan natives were also to supply labor and contract services to the pipeline project, and so upgrade skills among a people long living in near poverty on their own ancestral lands.

States' rights. Oil brings Alaska, youngest but one of the fifty states, into the battle over the rights of a state in a federal system, an issue as old as the United States itself. The pipeline project and now the oil royalties are restructuring Alaska's very economy, long dependent on government spending, tourism, and nonpetroleum resources. In just a decade, this sparsely peopled outpost has arrived on the threshold of wealth and the political clout that keeps company with wealth. Its power may rival other oil-rich U.S. states and even some of the nations of the Organization of Petroleum Exporting Countries (OPEC) whose actions drove western oilmen to challenge the chill realms of the Slope in the first place. Offshore, in the Gulf of Alaska, the state is already in confrontation with federal outer continental shelf leasing policy. And Alaskans love this. They are fiercely independent of the U.S. mainland they call the "Lower 48" or "Outside"—whose distant government let them languish even without full territorial rights from 1867 until statehood in 1959. Now Alaskans can talk back, and they are doing so.

The catalog could go on. Whatever the issue, though, you'd likely find some reflection of it in Alaska. Large forces, like Alaskan glaciers, grind and mold and slowly alter all that they encounter. Most of the large forces of history have come into play in Alaska over the centuries, and especially in recent years. The ten-year tale of the Alaska pipeline is a story of those forces and of their interplay with each other—of issues and the impacts of those issues.

More superlatives? Of course. Alaska is a land of superlatives. If this book is riven with them, it only apes the place where its story happens.

1.
1968 ... Is a Pipeline Possible?

Amid the crush of tourists pouring into Alaska each summer, the men from Texas drew little attention at the Anchorage airport that last week of July 1968. Pipeliner-fashion, hedgehopping across the thousands of miles from the humid Gulf Coast to breezy Alaska, they made up their own small planeload aboard one of Humble Pipe Line Company's Grumman Gulfstreams. The Humble Pipe Line engineers were Frank A. Therrell and Joseph V. Neeper, and with them was William S. Spangler, a company vice-president. A. F. "Buddy" Morel and Christopher B. Whorton were engineers from Atlantic Pipeline Company, with their vice-president, Richard G. Dulaney.

The small group had an assignment that not many people even in their own companies knew much about: to begin the first feasibility studies for the Alaska pipeline.

It was barely five months since the first hint that there might be enough oil under Alaska's North Slope for a pipeline to make sense. In mid-January, Atlantic Richfield had announced some tentative drilling results at Prudhoe Bay State No. 1, the exploratory well it had begun drilling in April the year before in a joint venture with Humble Oil & Refining Company. The well, ARCO said, had "returned a substantial flow of gas" at 8,500 feet. In February came reports of oil-saturated rock samples from the well's core. In March the flow test rates were more encouraging: 1,152 barrels of oil per day and 1.32 million cubic feet of natural gas at depths between 9,505 and 9,825 feet. Drilling would continue to 13,000 feet. Meanwhile, to test the size of the field, ARCO and Humble put a second drilling rig to work near the Sagavanirktok River, seven miles to the southeast.

In June the full extent of the find at Prudhoe Bay State No. 1 was made known. At 8,656 to 8,670 feet, the well tested up to 2,415 barrels of oil per day—a commercial discovery. Between 8,210 and 8,250 feet, the tests showed gas at a rate of 40 million cubic feet per day. The second well, Sag River State No. 1, had also found oil at the same depths, confirmed soon at a flow rate of 2,300 barrels per day. On the basis of the two wells, said the Dallas geological consultants DeGolyer & McNaughton: "This important discovery could develop into a field with recoverable reserves of some five to ten billion barrels of oil, which would rate as one of the largest petroleum accumulations known to the world today." Prudhoe Bay was the largest oil field ever discovered in the United States. It was also about one hundred times more oil than anyone had found anywhere in northern Alaska before.

On the New York Stock Exchange, the share prices of Atlantic Richfield and Standard Oil Company of New Jersey, Humble's parent corporation, went predictably berserk. So did those of any other company even rumored to hold Prudhoe Bay leases. And, far away at the London Stock Exchange, so did the shares of British Petroleum Ltd., which had just left the North Slope in 1967 after eight lonely years of looking unsuccessfully for oil on its leases there. Now, in early July, BP quickly sent a project team over the Pole to plan a new drilling program on its acreage for the next winter.

Meanwhile Humble's Frank Therrell and Joe Neeper and ARCO's Buddy Morel and Chris Whorton were all but airborne to Alaska by the time Prudhoe Bay's impressive oil potential burst on the world. As soon as the discovery was confirmed, ARCO and Humble had formed a crude oil pipeline task force. The field study team flew off to Alaska while more engineers waited in Texas to see what they found out. "We were already doing office exercises in Houston," Frank Therrell says today. Therrell, in fact, had been studying Arctic pipelines for two years. Early in 1966, as ARCO and Humble began drilling their first unsuccessful North Slope well, Therrell had just come back to Houston from an assignment in Italy. "They turned me around right away and sent me up to see L. A. Kenstock at Imperial Oil in Calgary, Canada. He was our most knowledgeable person on working in permafrost," Therrell recalls. Imperial Oil Ltd. was the Canadian affiliate of Jersey Standard, now Exxon Corporation.

Chris Whorton, from ARCO, had just been in Alaska in the spring of 1968. "I guess, when I came on the project, I was the only one who had ever done any work in Alaska," he says. "I had been over at Drift River on the west side of the Cook Inlet, where we've got an interest in some pipelines. I just got back to Dallas to start on some of the engineering and design concepts before the four of us came up here." As it turned out, Frank Therrell stayed on the Alaskan pipeline project longer than any other person, first as an engineering manager and then as manager of project permissions through the pipeline's completion. Chris Whorton left Alaska only in late 1976 after more than eight years on the engineering staff. Joe Neeper and Buddy Morel each spent four years in Alaska before their companies reassigned them to management jobs in the Lower 48.

In the summer of 1968, though, it was one day at a time. Summers are long in Texas, but brief in Alaska. It was to be a busy season for the pipeliners in both places. Several basic questions about an Alaska pipeline had to be answered as quickly as possible. What route should it take? Where should its tanker port be located? How much oil should it be built to carry at first, and how might it be expanded later? At what temperature should the oil be carried? How many pump stations would be needed to move the oil along the line, and how should they be powered? What kind of pipe should be used for a pipeline in Alaska, and where could it be obtained?

Some of these questions were for the engineers back at headquarters. Some could only be studied firsthand in the field. More men from Texas came to Alaska—consulting engineers from Pipeline Technologists, Inc., in Houston, with a contract to study possible routes and terminal sites. To oversee the expanding business and logistics of the effort in Alaska, the companies sent up accounting and purchasing managers. In mid-August, British Petroleum agreed with Humble and ARCO to share a third each of the study costs, and continued assigning its managers to new posts in Anchorage.

With half the summer already past, the field studies started immediately. Joe Neeper had come to Alaska from a job as Humble Pipe Line's district manager in Lafayette, Louisiana. Now he and ARCO's Buddy Morel were project managers and field engineers for their companies. "We had to arrange everything all at once . . . the routing studies, the soil studies, the port facility studies, how to get the field teams out, where to get helicopters, how to get living quarters," Neeper recalls. By the first week in August, they had five teams out in the field tramping over possible routes along the entire 800 miles from Prudhoe Bay south, and looking at several terminal sites on the northwest coast and along the Gulf of Alaska. From the beginning, the ground parties were supported from above by aerial surveyors.

That same week, Frank Therrell and Chris Whorton were on their way to Fairbanks. The ARCO district office in Anchorage had suggested they talk to a Dr. Harold R. Peyton at the University of Alaska. A civil and structural engineer, Hal Peyton by 1968 had spent more than a dozen years in Arctic research. He had lived at the North Slope and had been a consultant to ARCO, Humble, and other oil companies on design of drilling platforms and dock facilities in the Cook Inlet, one of Alaska's earlier oil centers. Now, as a consultant again, he was to be in the midst of the first fieldwork on the Alaska pipeline, and later would join the project as manager of staff engineering.

In Fairbanks, Peyton introduced the pipeliners to Ralph R. Migliaccio, a partner in R & M Geological Consultants, experts in soil mapping and engineering geology. Therrell and Whorton then continued as far north as they could go in Alaska, to Point Barrow on the shore of the Arctic Ocean. Here was the U.S. Naval Arctic Research Laboratory and its director, Dr. Max Brewer. The visit would set the groundwork for the project's first tests of what happens when you bury a pipeline in permafrost. As far as anyone knew, this had never been done before in the latitudes of Prudhoe Bay and its oil field.

There were many things about the proposed pipeline that hadn't been done before. The companies had spent years looking for oil on the North Slope. They knew the rigors of surveying and drilling in those climatic and soil conditions. But building a pipeline south from there, with more than a third of it above the Arctic Circle, would have a new magnitude of problems. As project planning began, a search was made for all available literature on Arctic engineering, construction, and pipelining. The best known "handbook" then on Arctic problems was the *Proceedings of the First International Conference on Permafrost*, held at Purdue University in the mid-1960s. There was the work of the naval laboratory at Barrow. Other sources were the U.S. Army's Cold Regions Research and Engineering Laboratory at Hanover, New Hampshire, and the Arctic Environmental Engineering Laboratory at the University of Alaska. All of these became consultants to the pipeline project. There had also been the construction experience gained as Alaska itself was built up as a defense bulwark during and after World War II.

But for all of this earlier research and experience, there was little of real use. "There was more work done in Arctic engineering than those of us who came up here expected. But there was not nearly as much as we would like to have seen," Chris Whorton says. Frank Therrell's 1966 trip to Canada only confirmed that most Canadian pipelining was in sub-Arctic muskeg regions. And, said Therrell, "When we had the Russian reports translated, most of it turned out to be theoretical . . . just some guy's idea of how it might be done." Hal Peyton summed it up: "Our work has used virtually zero Russian input. They had not built any facilities of this sort in this environment. And the Canadians had done precious little."

The Alaskan environment raised one key question that few of the world's major pipeline projects had ever needed to consider. What temperature should its oil be? Conventional pipelines in temperate climates transport their oil at its normal production or storage temperature. The Alaska pipeline was to be built largely through permafrost terrain—not just above the Arctic Circle, but over most of whatever route was picked. By the classic definition, permafrost is perennially frozen ground, remaining below the freezing point, in its natural state, for at least two consecutive years. It is by no means found only in the Arctic. But oil comes from the well at temperatures as high as 160°, and while it may cool off a bit in storage, it is heated up again by hydraulic friction as it moves through a pipeline and its pumping stations. The heat of the oil also warms up the steel pipe itself. So an early worry was that burying a warm pipeline in permafrost, as it would normally be done, would have more or less the same effect as trying to put a heated nail through an ice cube.

The task force looked first in 1968 at designing a cold oil pipeline, with the oil constantly chilled between 30° and 35°. Therrell, Whorton, Peyton, and Ralph Migliaccio made more trips to Barrow, where an unusual test facility was built and instrumented. A thousand-foot section of 40-inch pipe, with bends that went up, down, and sideways, was buried in varying depths of permafrost to test the interaction of pipe and ground. Pieces of smaller-diameter pipe were buried to analyze the stress effects of another Alaskan subsoil phenomenon: the ice wedge. What's an ice wedge? It's a vertical underground iceberg, formed over long periods of years by annual cycles of freezing and thawing. Ice wedges and their horizontal cousins, ice lenses, would put stress on a pipeline as they heaved and cracked during seasonal temperature changes. They might be found at any point on the pipeline route. And they have been. So has ice from long-departed glaciers. The Barrow test facilities also included several plots of tundra, the insulating ground cover that overlays permafrost. These were used to test how to use construction machinery with minimum damage to the terrain, and how to restore terrain that was damaged.

A cold oil pipeline seemed a good idea at first. But it turned out not to be, for both technical and cost reasons. For one thing, there would be too much oil coming from the field to store at Prudhoe Bay while it was chilled down. Large and expensive tank farms, each requiring some 100,000 horsepower in mechanical refrigeration, would loom up all over the North Slope landscape. Also, as discoveries on the Slope built up, it would make more sense to "unitize" the oil field and have each half of it operated by one company for the others who held leases there, and bringing the oil from the wells to the pipeline through common gathering systems.

The oil in a cold line would have to be chilled not only initially, but all along the route as friction heated it up again in passage. Each pump station would need more refrigeration equipment to rechill the oil. And two to three times as many pump stations would be needed to move the sluggish colder oil as for warm oil. Another cost would be chemical additives to maintain the cold oil at a pumpable density that would keep flowing, for there was a fear that if a cold oil line had to be shut down for any reason, the oil would become a sludgy mass hard to start moving again. But the biggest problem was that all crude oil contains wax. Chilling it causes the wax to separate out and be deposited on the walls of the pipe and storage tanks, requiring frequent use of mechanical "pig" scrapers—plus the need to dispose of some 1,400 tons of wax that would accumulate every day.

The technical problems of a cold oil line were bad enough, but each one also meant higher costs in some part of the pipeline system. Ultimately, it was found that the total cost of a cold oil line would be almost two-thirds higher than a hot oil line. Higher construction and operating costs would put too high a price on North Slope oil. So, as the 1968 studies went on, the engineers were having severe second thoughts about a refrigerated pipeline. The more conventional hot oil line was looking better.

The survey parties in the field, meanwhile, were looking at possible pipeline routes. The decision between cold and hot oil would be a factor in picking the final route. But the first job was to decide on a port location. A terminal on the northwest coast would have led to some complex Arctic engineering problems. The better plan was that the line should begin at Prudhoe Bay, where the oil was, and end somewhere at the state's southern coastline on the Gulf of Alaska. That meant a route of about 800 miles, running more or less north to south, depending on where the port was placed. Whatever the route, it would have to cross Alaska's three major east-west mountain chains: the high Brooks Range in the north, the lower Alaska Range to the south, and the Chugach Mountains at the coast.

This was not the first search for a pipeline route through Alaska. In World War II the Alaska Defense Command had a unit called the Alaska Scouts, made up of Alaskan natives and other American Indians. The U.S. Navy was just starting the first serious oil exploration on the North Slope, and the Scouts surveyed a pipeline route from Barrow to Fairbanks. In the early 1960s, when British Petroleum and other oil companies began probing the Slope, more pipeline studies were begun. This early work was the basis for new route explorations now.

The pipeliners looked first at two basic routes, each with several variations. Both were keyed to low maximum elevations. One alternative led up more or less from the Colville River on the west side of the North Slope to 2,400-foot Anaktuvuk Pass in the Brooks Range. The other would follow the Toolik River from the Slope to the same pass. But both would run the line through long regions of silt rich in troublesome ice deposits, Hal Peyton suggested looking at a route to the east. It would follow the floodplain of the Sagavanirktok River, an area of gravel deposits offering a stable pipeline bed. From there, this route followed the Atigun River to Atigun Pass on Alaska's continental divide. On the plus side, long stretches of icy silts could be avoided. The problem here was that Atigun, at 4,790 feet, was twice the altitude of Anaktuvuk Pass, and would pose great hydraulic problems of raising oil over those heights if the oil were being pumped cold.

The pipeline route could never hope to bypass all of the permafrost and other difficult soil conditions in its path. This raised a second major question: Would portions of the pipeline have to be elevated instead of buried? Even after brief study, the project's planners were already conceding that perhaps 5 to 10 percent of the mainline could not be buried. "The ice rich permafrost has not been avoided altogether," stated one of the earliest research summaries. "It appears that there will be approximately 40–50 miles of above-ground construction required along the selected route in relatively short, discontinuous lengths." That projection would prove conservative by a factor of ten.

From the continental divide, the alternate routes descended southward—along the John River from Anaktuvuk, or along the Koyukuk River from Atigun. Now the pipeline would cross the Arctic Circle and head for the fabled grandfather of all Alaskan rivers, the Yukon. Sub-Arctic or not, permafrost conditions were not improving. They were getting worse, and the route surveys were aiming to use as much thawed ground and bedrock as they could find. At this early stage, plans were to bury the line deep under the Yukon—a river that went on the rampage each spring during what Alaskans call "breakup," sending the winter's thawed ice crashing westward from Canada to the Bering Sea, and causing a lot of what river engineers term "scour" in its broad, meandering stream bed.

South of the Yukon, the pipeline route would now be in inhabited Alaska. Its direction the rest of the way depended on which port it was heading for. There were several candidates.

Whittier and Seward, on the Kenai Peninsula south of Anchorage, were established southern Alaskan ports. So was the shore of the Cook Inlet, where tankers already called regularly for cargoes of Alaska's only significant oil production to date. A possible site was Cordova, a fishing port at the gateway to Prince William Sound, a hundred miles or so east of Anchorage. And at the head of Prince William Sound was another fishing port, Valdez.

Valdez came to be favored on many counts. Except for the Cook Inlet, it was the farthest north and could be reached by the shortest and most direct route for the southward-bound pipeline. Valdez was more sheltered than the other sites and stayed free of ice all year. Seward was frequently exposed to heavy sea winds, and floating ice in the Cook Inlet would hamper year-round tanker operations. One caution was that Valdez had been heavily damaged by the earthquake and *tsunami* wave that had ravaged southern Alaska in 1964. So had Seward's port area. But an elevated bedrock terminal site was available at Valdez across from the rebuilt town, and the wind and tide conditions in the waterways leading to it looked favorable.

A pipeline route from the Yukon to Valdez would swing eastward past Fairbanks, more or less following Alaska's major all-weather road, the Richardson Highway. Now the route was rising again, toward the Alaska Range, but much more gently than its steep climb over the Brooks. Permafrost was still a problem, and the surveyors still hugged the ridgetops to take advantage of exposed bedrock there. This would minimize soil erosion that might result from traversing the hillsides. From Fairbanks, the tentative route was laid first along the Tanana River and then the Delta River Canyon, crossing the Alaska Range through Isabel Pass at 3,500 feet. Next it would either cross or skirt the east edge of the Copper River Basin to the Tonsina River, and follow several smaller rivers to 2,500-foot Thompson Pass in the Chugach Mountains. The final portion would descend through Keystone Canyon to Valdez.

As the field engineers sent back their early data on routes, ports, and permafrost to their companies, the Alaska pipeline was taking its first rough shape. So was an organization to design and build it. In late October TAPS—the Trans Alaska Pipeline System—was formed as a joint venture. Humble, Atlantic Richfield, and British Petroleum each held a one-third interest and began assigning more managers and engineers to the venture. George G. Hughes, Jr., of Humble Pipe Line was project manager, and Robert B. Hastie of ARCO and Samuel M. Corns of BP were assistant project managers. In November David Henderson arrived from BP in London for his first look at the project work. An engineering manager, he would return early the next year as resident manager for TAPS in Alaska.

The first design staff was put together under Vernon Cardin of Humble Pipe Line as manager of engineering. Hollie M. Childress of Humble was the mechanical design supervisor, with a coordinator for each project segment: John H. Shingle of Humble on the mainline, Walter McDermon of Humble on the pump stations, and Gordon Gouldston from BP on the terminal. Humble's Joseph L. Willing worked then on the system's hydraulics and would later join the project in Alaska. Robert B. Giezentanner of Humble was purchasing and logistics manager, and BP's Harry Heywood came from London as his assistant. F. W. Charlwood of BP was manager of estimates and cost control.

Vernon Cardin was on the project for four years before returning to Humble Pipe Line. "Each company had been doing its own study . . . the transportability of crude oil out of Alaska, whether or not we could build a pipeline, how much it would cost, and when we could get it done," he recalls. "Then we formed a group in the fall of 1968 to do an intercompany feasibility analysis and cost estimate." Cardin's job was coordinating the technical information that was pouring back from Alaska by phone, in writing, and often in person. Joe Neeper remembers making four round trips to Houston in one month on the route surveys. "My body didn't know what time it was," he says now.

Frank Therrell and others, at that same time, were sending or bringing back engineering data. There were also reports from each company on the characteristics of North Slope oil. "We could work on the hydraulics from that information," says Hollie Childress. "Our basic job was trying to 'size' the facilities and come up with some preliminary flow plans . . . pumping units, horsepower demands . . . and some cost estimates for the various pieces of the system."

The challenges of pipelining in Alaska were starting to assert themselves. Childress sums them up. "We had to look at the pumping conditions up there . . . the thermal effects of pumping oil in a large-diameter line," he says. "We had to get suitable low-temperature properties for all of the materials involved . . . the mainline valves, fittings, flanges, and traps. We had to house each facility adequately for the weather, and make each pump station self-sustaining. Then there was the remoteness of the location . . . you had to get everything there in pieces that could be transported. And as the project went on, there was the earthquake design, and the need to respond to environmental requirements."

But these last two factors were only beginning to loom up over the pipeline project. In Alaska the arrival of winter was closing down the work of the study teams in the field. In Houston the TAPS engineering group was on deadline to put out its first project report to the three owner companies. Now the six months of field studies in Alaska and engineering calculations in Texas were drawn together and boiled down into a small book with a gold cover—the best conclusions so far about how to build a pipeline from the Arctic. "The companies wanted a project outline . . . something with a reasonable cost, and quick off the ground," recalls Harry Heywood, now with BP Trading Ltd. in Harlow, England. "We finished the feasibility study on December 12, 1968 . . . in time for us from BP to do our Christmas shopping at Foley's department store in downtown Houston and come back to the U.K."

2.
1969...Poised to Start

In Anchorage, on February 10, 1969, Atlantic Richfield, British Petroleum, and Humble Oil announced the Alaska pipeline. It would be a 48-inch line running 800 miles from the North Slope to the Gulf of Alaska. Its initial capacity would be about 500,000 barrels a day. The exact route and terminal location were still being studied. Construction surveys would start that spring. Completion was expected in 1972. The total cost of the system was estimated to be about $900 million.

TAPS was to be owned through the pipeline subsidiaries of the three partners in the 1968 study, on a slightly different percentage basis. ARCO and British Petroleum now held 37½ percent each. and Humble held 25 percent. In the future, if other companies needed to bring out oil from the North Slope, they would be invited to share in the line's ownership. As Prudhoe Bay production built up, TAPS planned to add more pump stations to raise capacity from 500,000 barrels to 1.2 million barrels a day by 1975, and to a maximum design throughout of 2 million barrels a day by 1980, if enough additional oil was found at Prudhoe Bay. The companies also said that while an Alaska pipeline would solve the prob-

21

lem of getting North Slope oil to the U.S. west coast, studies of other routes and methods to bring the oil to the midwest and eastern U.S. were continuing.

The first construction task was not the pipeline at all. An all-weather supply road had to be laid almost from Fairbanks to the Slope. None existed. Alaska's public highway system stopped a few dozen miles north of Fairbanks. Even through 1968, as excitement over the oil strikes rose in Alaska, there was no way to reach any location from south of the Yukon to the Arctic Ocean unless you flew or walked. Over the winter, Alaska's development-minded governor, Walter J. Hickel, had tried to help things along by having the state build a road from Livengood, below the Yukon, across Anaktuvuk Pass to Prudhoe Bay. In this season the road could only be a winter trail crudely slashed through the frozen wilderness. It was slow and risky to travel. It was useful only until the spring thaw. Anyway, who had said Anaktuvuk would be the pipeline's final route? The "Hickel Highway," as it was derisively nicknamed, would have to be replaced by a properly engineered construction haul road, built from scratch, to get men, materials, and equipment to the entire northern half of the pipeline project.

TAPS planned to build that road in the summer of 1969. The pipeline construction force would be mobilized over the next winter, ready to begin as the weather improved in 1970. This construction schedule meant that the 800 miles of pipe itself had to start arriving in Alaska in the fall of 1969 so that the job of distributing it section by section over the final route could begin. Now all TAPS had to do was find 800 miles of 48-inch pipe.

The key here was the pipe's metallurgy. Within the Houston engineering group was a metallurgical task force: E. L. von Rosenberg of Humble, Rado Loncaric of ARCO, and Harry C. Cotton of BP. Hollie Childress also worked on specifying the pipe. The final pipe specifications were tough ones. Pipe to be welded and installed in Arctic and sub-Arctic Alaska had to have special low-temperature characteristics. This meant using a steel with a low carbon content. Other chemical impurities had to be restricted. The pipe also had to be unusually ductile—able to change shape slightly in temperature extremes without being permanently deformed. This meant the steel had to have a small grain size, and the oxygen used in modern steel-making must be totally purged from the finished steel. TAPS wanted pipe with two wall thicknesses: 0.462-inch and 0.562-inch. It asked for unusually close manufacturing tolerances both on wall thickness and roundness. The pipe ends were to be trimmed carefully to aid in welding the joints under the harsh field conditions expected. High yield strengths were required—the pipe was to be able to withstand maximum yield strengths of 70,000, 65,000, and 60,000 pounds per square inch, depending on wall thicknesses and other factors.

In a way, the toughest specification was diameter. When the first price inquiries were sent out, TAPS asked for quotes on 36-inch, 42-inch, and 48-inch pipe. But it was made clear that the largest size had the priority. By the late 1960s, no long-distance 48-inch pipelines had been built. The initial inquiries were sent to nine American steel companies and a number of others in Japan, England, Germany, and Italy. But no U.S. company yet had a plant that could make 48-inch pipe. In fact, no such plant would even be announced in the U.S. until mid-1976. Several companies, in 1969, said they would be willing to build a new mill, either in Alaska or elsewhere, to supply the pipeline. But none, even then, could have started deliveries for more than a year.

The metallurgical team, including Cotton, Childress, and Harry Heywood as the purchasing representative, went shopping abroad—first to Japan, then to Wales, then to several pipe mills in Germany. In February TAPS put out its formal request for bids, allowing a month for responses. The American companies, even with an added month of time to submit bids, still could meet neither the specifications nor the delivery schedule. The Japanese steel industry could. In April Sumitomo Metal Industries, Nippon Kokan Kabushiki Kaisha, and Yawata Iron & Steel Company received a joint contract to start delivering $100 million in pipe the following September. The price tag was impressive at the time. In later years it would stand as the project's most spectacular bargain purchase.

TAPS had said that the route and terminal studies were still going on. In mid-March, as winter let loose its grip on Alaska, the ground parties returned to the field to refine their 1968 surveys of possible pipeline routes. "Frank Therrell and I set up this little field office in an old studio in Fairbanks to expedite the work," Joe Neeper says. "Then we went out with a crew along the "Hickel Highway", carrying our living quarters with us. All the rest of the supplies came behind us by plane." In Alaska, as in so many other places, the Lockheed Hercules was the workhorse aircraft.

The route surveys were air-supported in many ways. Frank

Therrell hired the famed Alaskan bush pilot Don Sheldon to fly low over parts of the tentative route while Therrell untwirled a spindle of twine as a trail for the surveyors on the ground. Two airborne geophysical contractors were on the project now: Geotronics, Inc., a unit of Teledyne, Inc., in California, and Edgar Tobin Aerial Survey Company from San Antonio, Texas. New techniques for remote sensing of terrain details from the air were being tried, including side-looking and ground-probing radar, and air-to-ground thermal scanning.

On the ground, crews were making the first soils explorations over the two main northern route alternatives through the Anaktuvuk and Atigun passes. This meant boring holes two inches in diameter deep into the ground every few miles to take out core samples in much the same way geologists test an oil well being drilled. This spring, more than 350 soil borings, with several samples from each hole, were made from Fairbanks to Prudhoe Bay. R & M Geological Consultants in Fairbanks contracted to coordinate the aerial and ground surveys into a geophysical map of the route in all of its variations. Another engineering firm, Woodward-Lundgren & Associates—later, Woodward-Clyde—was working with Hal Peyton on the thermal problems of matching up a pipeline with Alaska's frozen slopes and valleys.

These firms, joined by Dames & Moore of San Francisco, were also analyzing Alaska's past earthquake history, particularly along the possible routes from the Yukon south. Their basic research was submitted to Dr. Nathan M. Newmark and his associates, William J. Hall and Alfred J. Hendron, Jr., at the University of Illinois. Newmark was the ranking U.S. expert on earthquake engineering, and had been a seismic consultant all around the world on large dams, tall buildings, and nuclear power plants.

By now, a few months after finishing their preliminary project study, the TAPS engineers were all but convinced that a cold oil line was impractical. In Fairbanks, a hot-test facility was built on the University of Alaska campus. Sections of 48-inch steel culvert were buried in an area combining both permafrost and loess—a wind-deposited soil found in many places along the route alternatives and known to be especially prone to erosion. Hot oil pipeline temperatures were simulated by using furnaces to blow hot air along the inside walls of the culvert sections, and the surrounding ground was instrumented to measure the degree of thawing. The surface of the test site was also planted with birch and aspen trees and other forms of Alaskan plant life to measure a hot oil pipeline's effects on nearby vegetation.

Similarly the engineers were zeroing in on Valdez as the final choice for the terminal site. An extensive marine environmental study of the harbor at Valdez was begun. Again, scientists from the University of Alaska—the Institute of Marine Science, the Institute of Water Resources, and the Institute or Arctic Biology—were called in as consultants. From a quiet fishing town with little industrial activity of any sort, Valdez was now a prime candidate to be one of the world's major tanker ports. Leading in from Prince William Sound, Valdez Arm was one of the long, narrow, fjordlike bays that knife deeply into the south Alaska coastline. The scientists were looking at how often and how effectively the freshwater sources that fed the harbor would minimize pollutants and replenish the harbor waters. They were taking a census of the bay's marine life, and how its commercial fishing businesses would coexist with terminal activities. They were adding to an existing 40-year history of wind and tidal patterns, and to the area's own earthquake history. The entire potential terminal site, across Valdez Arm from the town, was being probed from the bedrock under its elevated hillsides down to—and under—the waters where docks and tanker berths might be built. By early summer, Valdez was the clear choice for the terminal, and a contract was given to Fluor Ocean Services, Inc. of Los Angeles for engineering design work on the Valdez facilities.

In 1968 the few months of field studies had barely been enough time for the nuts-and-bolts matters of where and how a pipeline from the North Slope might be built. Now, in 1969, the environment in which it would be built came under study. Alaska's ecology is as vast and various as the land itself. And in many ways that were hard to come to terms with on a harsh and rugged frontier, the state's ecology was regarded by environmental groups to be as fragile and delicate as spun glass. The pipeline route would be a slender tracing on a map of Alaska's 586,412 square miles. But, the environmentalists cautioned, it would still cross a wildlife habitat that had scarcely been influenced by human activity in all of its history. So now a mammoth industrial construction project, extremely difficult of itself, came face to face with such matters as the nesting cycles of peregrine falcons, the feeding, migration, and

hibernation patterns of bear, caribou, and Dall sheep, the spawning habits of salmon and char—and the habitat itself: the hillsides,
streambeds, and plant life that sustained nature's entire Alaskan
society.

There were two main questions: What would the project do to
the land during the two or three years of construction? And what
had to be done to correct any damage? Joe Neeper recalls another
early question. "We didn't know what the field crews could expect
. . . were they going to be attacked by bears, or what?" he says.
"Each party was furnished with a '44 Magnum rifle and a Magnum
pistol. . . . I don't believe they were ever fired except in target
practice."

In any case, TAPS was no doubt the first pipeline venture with
its own environmentalists on staff or as consultants from the start.
In the spring of 1969 a broad ecological survey was begun of the
area felt to be most sensitive: the wilderness roughly from Fairbanks north to the Slope. The team leader was Bryan L. Sage, an
ecologist from British Petroleum. The team's mammologist was
Peter Elliott from the University of Lethbridge in Canada's province of Alberta. From Regina University in Saskatchewan came
Peter McCart, a freshwater biologist. William Mitchell was a University of Alaska botanist.

The vegetation and revegetation studies were led by Keith
Van Cleve, a biologist with the University of Alaska's Institute of
Arctic Biology. His team included plant physiologists and
ecologists. A consultant on several of the studies was Jerry Brown
from C.R.R.E.L.—the U.S. Army's Cold Regions Research and
Engineering Laboratory. Other consultants came from Stanford
University and San Diego State College. And an Atlantic Richfield
engineer was also taking what would turn out to be a long-term
interest in the project. Albert C. Condo was already in Alaska as
manager of products and applications for ARCO's chemical division
in 1969. "Our board chairman, Robert O. Anderson, had taken a
flight over the possible pipeline routes," Al Condo remembers. "I
soon got a call that we were going to be involved in revegetation
work." Condo was shortly to be loaned to the project as supervisor
of Arctic civil engineering, and later would become supervisor of
restoration engineering as the pipeline company neared the startup
of its operations.

Meanwhile, as the pipeline studies went on, the new oil seekers were arriving at the North Slope each season in larger numbers. Prudhoe Bay's population of exploration crews had been on the rise since mid-1968 when pipeline research was barely begun. Mobil Oil Corporate had announced then that it would start drilling late in the year, operating for itself and Phillips Petroleum Company. Texaco, Standard Oil Company of California, and a number of smaller companies had also started North Slope operations in late 1968.

British Petroleum, one of the North Slope's earliest pioneers, had come back to the quest in force. As far back as 1952, when the company made an extensive inventory of all of the places around the world where it might one day want to search for oil, BP geologists had tagged northern Alaska as a place where large oil reserves might logically be found. For more than forty years, BP's greatest strength as an oil giant had been based on its huge producing concessions in Iran. That source was nationalized by Iranian Premier Mossadegh in 1951, and while the production rights were later regained from Iran's new national oil company, BP was now intent on diversifying its worldwide oil supplies. In 1958 it made a joint exploration and marketing agreement with Sinclair Oil Corporation in the United States. That same year, Sinclair and other U.S. oil companies had acquired the first federal leases on the North Slope. By 1959 BP itself had leases or options to lease on 150,000 acres around the North Slope.

The search was launched from BP's Canadian base in Calgary. A reconnaissance of the Slope was done in 1960, and the intensive surveys began in early 1962. F. Geoffrey Larminie, Irish-born with degrees in geology and zoology, came in from the Middle East. David J. Oliver, a Scot, came to Calgary as chief geophysicist of a new company, BP Alaska. The first seismic studies were done in the foothills of the Brooks Range, where the Navy had found its oil after World War II. From there the geologic parties worked south into the Brooks, planting their camps on lakes where float planes could land, and then north down through the Slope's coastal plain. "It was so empty up there then," Geoff Larminie remembers. "The only things you heard were aircraft, and the only people you'd see were other geologists."

BP learned enough that first year to bring a drilling rig over from Canada in 1963, floating it down through the Mackenzie Delta, across the Beaufort Sea to Prudhoe Bay, and up the Colville River where it could be moved overland to the first drill site. Over the next year, BP and Sinclair drilled six unsuccessful wells in the Brooks foothills. Meanwhile, in 1964, Alaska allowed the oil companies to bid for the first state leases on the North Slope, around the Colville River. But Sinclair left the venture before the first Colville well was started, and BP drilled alone—again unsuccessfully. Now, in 1965, the state leased the first acreage at Prudhoe Bay itself. BP by this time was teamed up with Union Oil of California, which drilled a second Colville well, but another failure. In 1967 the state held another Prudhoe Bay lease sale, and BP added to its North Slope acreage. It could find no drilling partners, though, and its own Alaska exploration budget was exhausted. After years of unrewarded efforts BP suspended its North Slope operations.

By the quirks of the oil business, BP was pulling out of the Slope just as the real excitement was about to start. In the 1965 Prudhoe Bay sale, BP had been outbid on one particular lease block by a joint venture of Humble Oil and Richfield Oil Corporation of California. So had Phillips and Mobil. And so had a middling-sized firm from Philadelphia: Atlantic Refining Company. But in late 1965 Richfield was looking for a merger partner, and Atlantic was looking to expand nationally. The two joined to form Atlantic Richfield, and with the deal came that Prudhoe Bay lease block. The new ARCO, with the block's other owner, Humble, put down the first well the next year and, in 1967, after that well came up dry, began drilling what would be the North Slope discovery well.

When the Prudhoe Bay field was confirmed—by the second well drilled with a rig BP had left behind in 1967—BP came back to the Slope in a hurry. ARCO and Humble, after their find, had offered to buy out BP's Prudhoe Bay acreage. But BP knew now that its North Slope geology was good. It had just drilled in the wrong places. Garth Curtis of BP's exploration department was rushed to Alaska from the Middle East to head the new search, and Alistair Lindsay was sent from London—"quite literally on forty-eight hours notice," he says—to take charge of purchasing supplies. A rig was found in southern Alaska and barged all the way around through the Bering and Chukchi seas to Prudhoe Bay. It began drilling at the Putuligayuk River in late November of 1968. By March of 1969 BP had its own first North Slope discovery. And, through the leases it had accumulated for almost ten years, it could

28

now claim to hold roughly half of the Prudhoe Bay field's total reserves.

Mobil and Phillips also found their first North Slope oil in March. By summer, Standard of California and some smaller companies had discoveries too. In August the ownership of the Trans Alaska Pipeline System was expanded. Humble kept its 25 percent. ARCO and BP each reduced their shares from 37½ percent to 27½ percent. The remaining 20 percent was taken up by five new owners. Mobil now came in for 8½ percent. Phillips and Union Oil each took 3¼ percent. Amerada Hess was in for 3 percent. And Home Oil Company of Canada took 2 percent.

The state of Alaska now decided to hold a final lease sale at Prudhoe Bay. With oil fever raging in Alaska, the contrast with earlier sales was startling. The three sales together in 1964, 1965, and 1967 had netted lease bonuses to the state of only about $12 million, at an average of a trifling $12 an acre. Even the discovery tract had gone to Richfield and Humble in 1965 for $93 an acre—and was thought then to be wildly expensive. But now successful wells were being brought in one after another. The new sale was held in Anchorage on September 10, 1969. It seemed as though the whole worldwide oil industry had gathered for the event. A total of 179 separate 640-acre lease blocks were offered. By the end of the day, dozens of companies and individuals had bid successfully for 164 of the tracts. The state was wealthier by $900 million, at an average bonus price of $2,180 an acre. There hadn't been a bonanza like this in Alaska since the Gold Rush, seventy years ago.

On the pipeline project, events were also coming in quick succession. Three days after the lease sale, on September 13, the first freighter arrived at Valdez with pipe from the Japanese steel mills. Deliveries continued at Valdez, Seward, and Prudhoe Bay for the next two years. In all, some one hundred shiploads were needed, each carrying more than eight miles of pipe in 40- and 60-foot sections. Some rough math tells how many single pieces of pipe that was. About 630 miles of pipe were delivered in 40-foot sections—more than 83,000 of them. About half of these were stored at Valdez, where shops were later set up to weld them into 80-foot "double-joints" and wrap them with a protective coating before they were trucked out to the pipeline route. The rest of the 40-foot pipe went by train and truckload to another storage yard at

Fairbanks, where it was also later double-jointed and coated. Meanwhile, 165 miles of 60-foot sections—about 14,500 lengths of pipe—were being freightered to Tacoma, to be transferred to barges for a long voyage around western Alaska to yet another storage yard at Prudhoe.

Another event was taking place the breadth of the North American continent away. In August the S.S. *Manhattan* churned out of the Delaware River near Philadelphia, bound for the top of the world. The 115,000-ton oil tanker, 1,005 feet long, had been taken apart in sections and rebuilt into a highly advanced Arctic cargo ship: a partially double-hulled tanker that was also an icebreaker. The $50-million research program, sponsored by Humble with support from ARCO and British Petroleum, was to see if regular tanker trips were possible in the Northwest Passage across the Canadian and Alaskan Arctic. The Canadian government was also lending research support to the program. Canada's coastal service icebreakers, and some from the U.S., were accompanying the *Manhattan* on its 10,000-mile round trip.

The voyages of the *Manhattan*—a second trip was made in the spring of 1970—were widely publicized. But they weren't completely understood. This was not a contest to choose between tankers and an overland pipeline. Oil cargoes through the Northwest Passage might be a second way of bringing Arctic oil to market— from Alaska, from Canada's developing Mackenzie Delta, from other prospective oil areas across the Canadian north. ARCO/ Humble's early task force in Alaska had also considered a possible tanker port at Wainwright, about one hundred miles west of Barrow on the state's far north coast. No route or method was being overlooked. But it was not going to be an "either-or" decision.

The *Manhattan* program was announced in December 1968. By that time, TAPS had already been formed, and the first project feasibility study had been finished. It was clear already that a pipeline could be built in Alaska, whatever other ways to bring out North Slope oil might be found. The pipeline itself would be announced two months later, long before the *Manhattan* sailed. The pipe order was also announced soon. Deliveries would begin within weeks of the tanker's sailing date, and the pipe's $100-million cost hardly looked like a speculative expense.

The overall costs of the two transport systems also had to be compared. On a cost basis, in 1969, they were rated as within range of each other. Just how close was hard to say. The TAPS owner companies were already acknowledging that the original $900-million estimate was well under the mark. The figure was soon to be raised to $1.3 billion, and some felt even then that $1.7 billion was more realistic. By contrast, the total cost of a Northwest Passage tanker fleet and offshore loading facilities at Prudhoe Bay was then being estimated at $2 billion to $3 billion. Technical factors and inflation would have boosted that figure, too.

But how successful were the *Manhattan's* pioneering voyages, after all? The Northwest Passage's ancient ice floes proved to be formidable adversaries. Time and again on the first voyage, the big tanker was trapped in encircling ice and had to be freed by Canada's icebreaker, the *John A. MacDonald.* Initially bound all the way around Alaska, the *Manhattan* struggled just to reach Prudhoe Bay. It took aboard a ceremonial drum of North Slope oil, went on the short distance to Point Barrow, and then headed back for the Atlantic. On the return trip, there was some ominous damage. Trapped again in the pack ice, the tanker was backing astern when a projection of steely ice ripped a long gash in a single-hulled section of the ship below the waterline. The *Manhattan* finished this trip and its second voyage safely. Future Northwest Passage tankers in any case would have had completely double-walled hulls against such danger. But there were severe doubts now. An oil spill in the Arctic would be catastrophic, for the oil would just stay there without decomposing for years. Also, the constant uncertainty of offshore ice conditions raised questions about the year-round loading capability that North Slope production would require.

Capt. Ralph Maybourn, now with BP Tanker Company Ltd. in London, was a technical monitor for BP on the voyages. "People have tended to write the *Manhattan* project off largely based on the *Manhattan* herself," says Maybourn. "She was the best available ship at the time . . . but she was clearly not suitable for the Northwest Passage." Still, he points out: "There was no damage whatsoever to the parts of the ship that were doubled-hulled . . . nor to the bow, which broke ice all through the voyage." The bigger problem, Maybourn says, is offshore loading: "whether," he says, "you can construct an offshore terminal to load tankers in a moving ice regime. I think it can be done, but what the cost would be is another question."

The pipeline, in any case was not a question. From the start of 1969 on, it was a committed project.

To build a pipeline, wherever you build it, permits are needed. You need a right-of-way to cross land, and permission to cross highways and streams and the like. Pipeliners had seldom run into problems getting permits—you applied, and they were granted. Alaska might have been expected to be a little different. To build a pipeline through a virgin wilderness—with a new road needed even to start construction, and a new terminal to be built from bedrock up—would likely require an unusual number of permits. Even before Humble and ARCO sent their first engineers to Alaska in 1968, they had requested and gotten permits from the U.S. Department of the Interior to start the field surveys. Later, TAPS asked for land clearances at Valdez so that it could get on with its terminal study. These, it turned out, were not going to be granted so quickly. What was the problem?

There were two problems. One was that Alaska was almost entirely public land. As a territory until 1959, of course, it was all federal land. The Alaska Statehood Act had given the new state the right to select 103.5-million acres—more than a fourth of its total area—as state lands. But this had not yet been done.

The other problem was related to this. The statehood law barred state selection of any land claimed historically by Alaska's Eskimos, Indians, and Aleuts. Again, little was done at first to resolve whose land was whose—until the oil prospectors began poking around the North Slope. As one state lease sale followed another, native leaders protested to the federal government. In 1966 Interior Secretary Stewart L. Udall had put a freeze on transfers of federal lands to the state and on any federal oil and gas leasing whatever, until the native land claims were settled. This freeze seemed to stand in the way of permits at Valdez, where the terminal site belonged to the U.S. Department of Agriculture. Interior asked its Bureau of Land Management and its Bureau of Indian Affairs to look into just what native claims did exist around Valdez. TAPS was assured that the delay would not be long.

In January 1969 the Nixon administration came into office. Secretary Udall, literally as his job expired, extended the land freeze to the end of 1970. He was succeeded as Interior Secretary by Walter Hickel, the governor of—of all places— Alaska. There was some tradition in naming a western governor to the Interior post. But there was also a budding new environmental spirit in the

Congress. Hickel was coming to the Nixon cabinet as the very model of the entrepreneurial Alaskan "boomer." In 1937, the Hickel legend has it, he had hitchhiked into Alaska with thirty-seven cents in his pocket, which he then bootstrapped into a multi-million-dollar real estate and construction fortune—and, in 1968, the state's first Republican governorship. Hickel's environmental credentials for the Interior secretaryship were sharply challenged during his Senate confirmation hearings. He assured the senators he would not remove his predecessor's land freeze as an obstacle to the pipeline in his home state without consultation with Congress.

To the amazement of all, Hickel was as good as his word. Interior first set up a North Slope Task Force, which was soon renamed the Task Force on Environmental Protection of Arctic Alaska. Its chairman was Interior Undersecretary Russell E. Train, who had headed a public interest group called the Conservation Foundation before taking office. The U.S. Senate Interior and Insular Affairs Committee, chaired by Senator Henry M. Jackson of Washington, appointed its own conservation committee.

In June TAPS applied formally for a pipeline right-of-way and for a right-of-way and permit to build the haul road. That same day, Russell Train submitted to TAPS a list of seventy-nine questions on technical and environmental aspects of the project. This was the first faint rising of the wind—an early sign of how detailed the government's scrutiny of the project was going to be. TAPS sent back its answers the next week. In late June, still taking his measured approach to approval of the pipeline, Hickel in a letter to TAPS said that permits would be granted "as expeditiously as possible," once Interior, the Congress, and the appropriate federal and state agencies were satisfied that all the legal, regulatory, and environmental requirements had been met. And as soon as "the interests of the Native people have been safeguarded. . . ."

In August, after notifying the Senate and House interior committees, Hickel did give the go ahead that allowed TAPS to build the first small section of the haul road. TAPS let a contract to a Fairbanks construction firm, and the work was finished early that fall. The road ran from Livengood, north of Fairbanks, to the south bank of the Yukon. Only summer roads had reached Alaska's great river before. The new road, unlike the Interior secretary's own "Hickel Highway," was a well-engineered, all-weather secondary highway, built above the terrain on insulating gravel pads. It would be a model later for the complete 360-mile road to Prudhoe Bay. It also gave TAPS its first extended lesson in Alaskan construction conditions. Joe Neeper watched the building of the Livengood road. "We'd cut five feet into a hill and find solid blue ice," he says. "That was a clue."

In August, too, the Interior Department issued a draft set of technical and environmental stipulations to accompany the pending pipeline permit. Two days of hearings were then held in Fairbanks, followed in September by TAPS presentations to both the Senate and House Interior committees. Alaskans and environmentalists alike had ample chances to give their opinions. Often these were the same people, for many Alaskan conservationists were against intrusions into the state's wilderness. Nationally, environmentalists had opposed the pipeline from the start. David Brower, the longtime combative head of the Sierra Club and now the founder of FOE—Friends of the Earth—testified to the Senate that the construction stipulations had loopholes "you could float the *Manhattan* through."

Development-minded Alaskans, on the other hand, wanted to get on with it. Mike Gravel, Alaska's Democratic senator, argued that oil development and the pipeline would raise the standard of living for natives in northern Alaska. He criticized the environmentalists as "those who want to put these people in living museums." And some Alaskans, voicing a longtime gripe, just wanted the people from Outside to mind their own business. The Fairbanks *Daily News-Miner* titled an editorial: "Alaska Capable of Preserving Its Beauty Without Interference." Said the *News-Miner:* "We know of no other state which has been subjected to so much 'protection' as has Alaska."

Conservation . . . environmental protection . . . native land claims. As the end of 1969 neared, instead of a handful of the usual construction permits, the pipeliners found themselves at the vortex of a growing handful of legal, social, and environmental issues. The original schedule had already slipped by a year. Now TAPS planned to gather its haul road work force this winter, ready to build the supply route in 1970, and start on the pipeline itself in 1971. Opinion was still divided among government staff members whether enough safeguards were being built into the Alaska project. Interior's top two men, Hickel and Train, seemed to feel that enough had been done to let the road work start. Remaining problems would be worked out before pipelaying began.

In mid-December work did get under way on a 2,250-foot ice bridge across the Yukon. This was a proven way to cross Alaskan rivers in winter. Layers of logs were put on top of the river's natural ice. River water was then pumped over the logs to form an added six to eight feet of ice that could support trucks and construction vehicles. Doing this work was Tundra Contractors, Inc. of Fairbanks, one of the first Alaskan native companies to work on the project. As the ice bridge was being built, TAPS on December 29 filed updated applications for both the pipeline and haul road right-of-way permits.

The TAPS staff in Anchorage was still expanding. Clyde O. Johnson of Humble Pipe Line had arrived fresh from an assignment in Jamaica. He was to be the project's manager of cost engineering, then its manager of project reports and, *ex officio*, its management historian. "When I came in 1969," Johnson remembers, "it was pretty much felt the permit was only six weeks away."

But Alaska, after the dramatic news of the North Slope, was no longer quite the little known, faraway wilderness it had been for centuries to most of the rest of the world. Now it had a new quality that reached well beyond its borders . . . "Alaska-Consciousness."

3.
"Alaska Consciousness" ...
A Sense of the Place

Polar Airways' twin-engined, eight-passenger Piper Navajo climbed doggedly through the misting drizzle above downtown Anchorage's Merrill Field. The pilot banked over the shrouded city and set a course eastward along Turnagain Arm, the long coastal fjord that almost cuts off the Kenai Peninsula from mainland Alaska. Somewhere about an hour ahead in the greyness was Valdez, southern terminal of the Alaska pipeline. "We'll get in there if there's a way to do it," the pilot called back to his passengers, who had already waited out three canceled flights this morning. "If we can't, we'll just have to land you at Gulkana and you can try to make it in by road from there."

Flying low across Prince William Sound into Valdez was worth the trip whether you could land or not. The Piper churned between Montague Island to port and Hinchinbrook Island to the starboard, above the entrance channel the tankers now travel from Alaska to the Lower 48. The craggy ankles of the Chugach Mountains rose steeply into the clouds that also hid Cordova, which might have been the pipeline's port. The plane entered the sheltered niche of Valdez Arm, so low that you could look up at the frontal overhang of a nearby glacier. A tourist and a pipeline worker unlimbered their cameras and clicked away. Despite the mist, rain, and clouds, the Piper landed lightly. Before it acquired a slick new terminal building in 1976, the Valdez airfield was a relic of the barnstorming twenties, with board shacks and pieces of junked airplanes dumped here and there—except that two of the shacks were the local Avis and National rent-a-car franchises doing an active modern-day business.

The hour's flight from Anchorage to Valdez covered less than two hundred miles. This far north, that's about four degrees of longitude. But even in that short distance, you've gotten a sense of Alaska's robust coastline and the ever-present challenge of its weather. In fact, though, you've barely seen Alaska's real distances. The pipeline's 800-mile, north-to-south route could easily fit within Texas or California. Luckily, it doesn't have to stretch west to east. From Attu Island, at the far western tip of the Aleutian chain, to Ketchikan, Alaska's gateway city at the lower end of the southeastern panhandle, is a distance of about 2,300 miles and some forty degrees of longitude. Alaskans can boast that the state has 33,000 miles of coastline—the coast is so broken up everywhere that arriving at its true length is like calculating the length of a saw by measuring up and down each of its teeth. Nonetheless, from west to east, Alaska spans four time zones, compared to the three that cover the entire Lower 48 states from California to Maine. If the International Date Line didn't cut sharply westward around Attu, a good portion of the Aleutians would be a day out of phase with the rest of the United States.

Alaska: a land of superlatives. It's a writer's dream—whether the subject is travel, nature, hunting and fishing, oil and other resources, social issues, or whatever. Alaska lends itself to adjectives in streams as undisciplined as its own rivers. Yet all of the adjectives will very nearly be accurate.

Alaska is a state more than twice the size of Texas. Its 586,412 square miles, or 375.3 million acres if you prefer, could contain Texas, California, and Montana within its borders. The entire state lies farther north than any other part of the United States, with more than a fourth of its land above the Arctic Circle. Alaska rises higher in altitude than any other part of the U.S. Mount McKinley in the Alaska Range, at 20,230 feet, is almost half again as high as California's Mount Whitney.

Alaska is also a land of sharp paradoxes. For all of its bound-lessness, the state has a population of about a third of a million people—roughly that of Rhode Island or the city of Newark, New Jersey. Alaska's population density is not much more than a half a person per square mile. If Manhattan Island were as thinly settled, about a dozen people would live there. Alaska, since it became a state, has rated the usual two U.S. senators. But its sparse citizenry calls for only one lone congressman in the House of Representatives.

The economy of Alaska is shot through with contrasts like this. Statistically, Alaskans have the highest per capita income of any state. Recently, this has meant an average income of more than $3,000 per person. This was so years before the high-paying pipeline project came to the High North. But Alaska, by its remoteness, also has high living costs that largely offset its income levels.

Alaska has many of the other typical distortions of a frontier economy. Its businessmen and high proportion of opportunity-seeking entrepreneurs earn well. Its large number of government people—military and civil service—are also well paid. But at the low end of the scale, Alaskan natives—who are 19 percent of the population—for the most part have lived barely at the subsistence level. A lot of older Alaskans are also on fixed incomes that do not rise with the state's inflationary economy, another frontier characteristic. Even as its average income figures keep on outshining the rest of the country, Alaska often leads the U.S. in unemployment rates. All of this had been further aggravated by the state's severe weather—and by the chronic economic boom-and-bust cycles that have long been the benchmarks of the Alaskan saga.

Alaska is an ancient land. But its recorded history didn't begin until the 1700s, more than a century after Europeans settled eastern America. Alaska's natives were there from the dimness of unknown times. They came as three distinct ethnic groups, many believe, from other lands. Arctic Russia, Alaska, and northern Canada have much more in common with each other even today than any of them has with the rest of the modern nation it belongs to. The Bering Strait between Alaska and Russia, barely fifty miles wide, may once have been solid land. If not, it was frozen and could be easily walked across in winter, or sailed across in summer. Siberian Eskimos migrated eastward to northern Alaska and Canada, and still visited between the Asian and North American continents until 1926 when Soviet Russia began buttoning up its borders.

The Aleutian Islands also stretched most of the way westward toward Russia's Kamchatka Peninsula. Probably they were also settled from Asia. The Aleuts live some of the most rigorous lives of all Alaskans, out along their thousand-mile chain of great and small islands. It was the Aleuts who gave Alaska its first name: Alyeska.

Alaska's Indians may have come there from the east rather than the west—from the heart of North America and not directly across the Pacific. They settled in the interior and along the southern and southeastern coasts. Each of the three native groups, over the centuries, took widely separate parts of the great land for itself: the Eskimos in isolation in the north, the Aleuts far to the southwest, the Indians to the southeast. There was not much mingling of their cultures. In fact, where they did meet, especially in the south, warfare often broke out. Indians and Aleuts were in bitter combat as recently as 1942, when the Japanese invasion of the Aleutians united them against a greater enemy.

Somewhere in antiquity, though, there were broad ethnic bonds. Alaska's Indians may have come the long way around. But they may also share a common Asian origin with the Eskimos and Aleuts. The Norwegian explorer Thor Heyerdahl has traced the migrations of early peoples around the world. "Many observers have maintained," he recounts the theory in *Kon-Tiki*, "that the great Indian civilizations, from the Aztecs in Mexico to the Incas in Peru, were inspired by sporadic intruders from over the seas in the east, while all the American Indians in general are Asiatic hunting and fishing people who in the course of twenty thousand years or more trickled into America from Siberia."

Almost 250 years ago, a Danish explorer also trickled into Alaska from Siberia. Vitus Bering was a ship captain from Copenhagen. But he sailed his expeditions for the Czar of all the Russias, Peter the Great. Russia's own vastnesses had only recently been politically unified all the way to the Pacific. The Czar was an imperialist looking for new lands and their riches. He also shared a European curiosity: Was there a northern land link between Asia and North America? He sent Bering six thousand miles overland to Siberia in 1725 to build ships and find out.

After three years of preparations, in 1728, Bering sailed into the wide northern Pacific sea later named after him. He discovered and named St. Lawrence Island, and explored enough of the Bering Strait to decide for himself that the continents were not connected. His Russian masters in St. Petersburg wanted more proof. In 1741 Bering led a larger expedition of two ships from Siberia. They took a more southerly course across the Pacific, and this time there were scientists aboard. Foggy weather separated the ships, and it was Bering's lieutenant commanding the second vessel, Alexei Chirikov, who actually first sighted the outer islands of southeastern Alaska on July 15. Two landing parties were sent ashore near Cape Addington—and both disappeared. Natives in canoes were seen. Fearful, low on fresh water, and with no more small boats aboard, Chirikov made the decision to sail back to Siberia with no real knowledge about his discovery.

Bering's own ship was farther north on the Gulf of Alaska. On July 16, sailing eastward of Prince William Sound, his party sighted Mount Saint Elias soaring 18,000 feet up in the Chugach Range. A brief landing for water was made on Kayak Island. Hastily one of the scientists aboard, the German naturalist Georg Wilhelm Steller, gathered plant and bird specimens, including a bluejay native only to the Western Hemisphere that would finally prove Bering had discovered northeastern America. The late Alaskan territorial governor, U.S. Senator, and historian, Ernest Gruening, called Bering's expedition "the last great voyage of discovery on earth."

It was Bering's last voyage, and he never completed it. Beset by scurvy with many of his crew, and disinterested in the pleas of the scientists to stay and learn more, Bering also headed home. Sailing past the Aleutian chain, he made a landfall on what was thought to be the Kamchatka peninsula north of his home port. The ship was wrecked in a storm, and the landfall turned out to be an island—later named Bering Island—due west of the Aleutians. Here Bering died in December, and thirty-one of his crew would also die. The survivors wintered on the island and returned to Kamchatka that spring in a boat built from the timbers of Bering's ship. Besides Steller's scientific samples, they brought back sea otter pelts from the island—the riches that had drawn them toward Alaska in the first place.

Russia claimed "Russian America" for itself. But Peter the Great had died, and so had Russia's interest in the scientific value of its discovery. For years, Russia did not even tell the rest of the world of the new land. When the news finally got out, Britain, France, and Spain all sent expeditions to Alaska. The British Admiralty in 1776 sent Capt. James Cook in search of the long-suspected

Northwest Passage between the world's two major oceans. But Cook also had secret instructions to claim for Britain any unsettled lands he found. In 1778 Cook sailed up the Alaskan inlet that would be named for him and into the branch that the sailors called Turnagain Arm. Neither, disappointingly, led through to the fabled transcontinental passage. But Cook laid claim to every landfall his expedition made, including one on Kayak Island where Bering had been. He then sailed southward down the long Pacific to discover, in 1779, the Sandwich Islands, later to be called Hawaii. In successive years, Cook had visited what would be the last two U.S. states. And Cook himself died at the hands of angry Sandwich Island natives. The 1700s were mortal years for the world's great explorers.

Russia did not settle Russian America for forty years yet. From the Aleutians eastward, Russian hunting expeditions brutally began harvesting any animal life that wore fur or feathers, at the same time slaughtering or enslaving the Alaskan natives as they went along. In 1784, at last, a town grew up on Kodiak Island near the neck of the Aleutian chain to serve as a center for the hunting and trading activities of the Russian-American Company. In the early 1800s the company's manager, Alexander Baranov, moved his base east to the Alaskan panhandle. From this settlement, New Archangel, the flamboyant, alcoholic Baranov ruled for almost two decades as governor, trader, and guardian of Russian interests throughout the upper reaches of western North America. The high life in New Archangel gained this remote village an unlikely reputation as the "Paris of the Pacific" among the sea captains of the era, and Baranov even "colonized" his realm as far south as Fort Ross, just above the Spanish settlements in California at San Francisco.

Baranov's great competitors in the Pacific Northwest were the British. The Hudson's Bay Company had been the model for his own trading enterprises. Long after Baranov's death in 1818, with diplomacy if possible and with warships if not, the Russians and British vied for commercial advantage throughout Alaska. Then, in the 1850s, far away on the Crimean peninsula by the shores of the Black Sea, Russia lost a war to Great Britain. The centuries-long Russian dreams of worldwide imperialism were shattered, and the court of Czar Alexander II decided to sell Russian America.

For reasons both of diplomacy and of geography, the United States became the favored buyer. Secret talks between the American Secretary of State, William H. Seward, and the Russian minister to the U.S., Baron Eduoard Stoeckel, led to a treaty of cession that was signed just before dawn on March 30, 1867. The agreed price was $7.2 million. On April 9 the U.S. Senate ratified the treaty by one vote. The narrow margin was an omen. Over the next century or so, Alaska would somehow inspire a great many closely contested votes on matters affecting its future.

In the United States of the 1860s, the few people who had actually been to Alaska—sea captains and military men—were enthusiastic about its acquisition. The rest of the country was appalled. Nicknames like "Walrussia" and "Icebergia" were coined for the unknown land. Alaska was described by people who had never seen it as a barren and inaccessible wasteland frozen solid to a depth of six feet. Congress delayed approval of the purchase terms for more than a year. Meanwhile, it had impeached President Andrew Johnson. The main issues of that came out of the aftermath of the Civil War, but Johnson's judgment in agreeing to pay an outrageous sum like $7 million for a half million square miles of "perpetual snow" also came up in the impeachment debate. Congress had kept Alaska by one vote. It also kept Andrew Johnson by one vote.

New lands absorbed in America's westward march normally became territories with an organized government and well-defined rights and privileges for their residents and anyone who later settled there. Not so Alaska. For seventeen years, it was first a military district and then a customs district. Its base of government was the old site of New Archangel, now called by its Indian name: Sitka. But most of the amenities even of a new frontier were missing. Lawlessness was typically widespread, for no courts existed to deal with it. The few troops stationed briefly in Alaska were dispatched back to Idaho in 1877 to put down an Indian revolt. The collector of customs pleaded to Washington for help. None arrived for almost a year, and then it was a sympathetic British ship. In the next years, the "governor" of Alaska was the captain of whatever U.S. Navy ship first arrived in Sitka each year.

Now Congress passed the Organic Act of 1884. It gave Alaska a bona fide governor, but little else. Civil structures like courts and a land office were defined but could not be put in effect. Alaska was to be governed by the civil code of Oregon. But homesteading claims could not be filed because Alaska had no land laws of its own. Courts could not appoint legal juries because Alaskan residents were not taxpayers and jurors must be taxpayers. Prohibition was declared in Alaska, which hardly dented the flourishing trade in moonshine because there were hardly any law enforcement officers to be found. Yet Alaska by now had a population, estimated in its first rough census, of almost 50,000 people. Most were still the natives, for whom more than two hundred years of Western civilization had brought little but exploitation, disease, and violence.

Alaska was about to become better known to the world. In the autumn of 1898, gold was found in a creek bed in the Canadian Yukon. The creek flowed into the Klondike River, which flowed into the Yukon River at Dawson. By the next summer tens of thousands of prospectors were laboring up Alaska's southeastern Chilkoot Pass from Skagway, and up the Yukon itself from the Bering Sea. More gold was found at Nome on the shores of the Bering, and in the Tanana Valley, where the frontier town of Fairbanks began to grow.

Much as the fur trade had lured the Russians, much as the California Gold Rush fifty years earlier had sped up that state's history, Alaska now had new commercial value. Congress, in its way, ground slowly into action. The Homestead Act was extended to Alaska. Geologic surveys began to reveal Alaska's other resources: coal, copper, and other minerals. Plans were formed to build roads and railroads. Congress took its time. Alaska was redesignated a territory, but in fact it was not. Not until 1912 was Alaska allowed any self-government, and even then many of the powers of territorial government were still withheld. The territory had little say in its own finances or judicial system, or in the use of its land. It did, at least, have a nonvoting delegate to the U.S. Congress. And it had its own legislature. The first act of the new Alaskan assembly in 1913 was to give the vote to the territory's female citizens—a provision not added to the U.S. Constitution until 1920.

But Alaska's notoriety beyond its own borders lasted about as long as the rip-roaring years of the Klondike and Yukon gold fever. Commercial mining of both gold and copper continued in the territory, but the excitement was mainly for the private mine owners. After some speculative rail ventures had failed, the United States government built Alaska's first and only railroad, and still owns all 470 miles of it. The push toward full territorial status—and beyond that, statehood—had resumed its glacial pace. In 1916 Alaska's delegate James Wickersham, who was an active and creative proposer of new laws to help the territory even though he could never vote on any of them, introduced the first bill to make Alaska a state. It would be forty-three more years and dozens more bills before Alaska actually got its state flag.

Alaska stagnated in the 1920s. Population had actually been declining slightly since 1910. This trend was sharply reversed in the 1930s. Despite the depression—or more probably, because of it—Alaska began attracting emigrants from the Lower 48. New Deal programs helped. The colonization of the Matanuska Valley north of Anchorage proved to skeptics that Alaska could in fact sustain flourishing farms in the midst of its "perpetual snow"—the almost around-the-clock daylight of the southern Alaska summer made up for the short growing season. And make-work programs like the WPA and CCC added valuably to the territory's port and tourism facilities.

World War II again put Alaska on the modern map. In late 1941, despite a decade of warnings as Japanese imperialism had spread through eastern Asia, Alaska had virtually no military defenses. Yet the Aleutians were only 650 miles from the northernmost Japanese islands. In June 1942, to divert attention from their attack on Midway and also because they feared bomber attacks from the Aleutians, the Japanese invaded and occupied Attu and Kiska islands without a casualty. They stayed for a year while the U.S. belatedly scrambled to build up its Alaska military forces. Attu was won back in May 1943, only after a bloody three-week battle.

Alaska hasn't been left unguarded since. By the end of the war, it was bristling with armament. It had air service from the rest of the U.S.—and a rugged road connection, the Al-Can Highway through Canada. During the Cold War, its closeness to Soviet Russia across the Bering Strait and over the Pole brought almost continuous construction activity on missile detection networks in Arctic Alaska—the "DEW-Line" and "White Alice" distant early warning systems. Alaska, on the Great Circle route across the Pacific, was also a staging area and stopover point during the Korean War.

War and its aftermath had done more than stress the territory's strategic value. Alaska had been popularized a little. Thousands of servicemen had seen the territory at its best and worst and came back to live there. Alaska's federal establishment was growing in proportion to a new perception that the place was "important—and many government workers also came and stayed, or came back later. By the 1950s, Alaska, where settlement had come hard, had nearly 200,000 residents.

Statehood was also coming hard. Postwar Alaska was a maturing frontier, partly emerging into the fast-paced contemporary world. But it was still mainly its own special place. The majority of Alaskans, at the war's end, seemed to favor statehood. But many did not. The same was so of the newcomers. Many were bullish entrepreneurs who wanted their new base of operations to come fully into the mainstream. But many had come to Alaska for its very privacy and remoteness, and wanted to keep it that way. Alaska was developing its "modern" personality—one of sharp splits and contrasts, riven with contradictions.

Within and without the territory, these opposed viewpoints clashed over statehood. Business interests that viewed themselves as progressive pushed for the full rights and privileges of being a state. But some Alaskan industries felt they were better off under the federal wing. Alaska, they said, was not mature or developed enough yet to take on the costs and burdens of statehood. Far away, in the Congress, the national politicians played cautious hands. The causes of Alaskan and Hawaiian statehood were inevitably paired. Would the new states be Democrat or Republican? How would their new representatives tip the postwar balance of Congress, especially the Senate, during years when the presidency and the congressional majorities were often of opposing parties? Even after almost a century, too, many national leaders still had the old ignorance about Alaska as a strange, almost uninhabitable, not very "American" kind of place.

Statehood came. The final bills passed both houses of Congress in mid-1958, by a comfortable but not overwhelming majority in the House of Representatives, by a 64 to 20 margin in the Senate. Alaska entered the union on January 3, 1959, as the forty-ninth state. Hawaii became the fiftieth state the next August. The two disparate places, separated by a thousand miles of ocean and worlds apart in climate and modern economic profiles, still share an affinity that goes beyond even their similar Pacific Basin cultures and trading histories. They are frontiers in common, with harsh heritages.

Alaska has remained a very private place even since statehood. It is still known best—perhaps known *only*—by its own people. They are there for more than the usual reasons that people wind up in various places. Alaskans still divide themselves into "sourdoughs"—the old-timers—and "cheechakos"—the newcomers. Both groups are distinctive. The sourdoughs have stayed because they wanted to. The newcomers have come to have a look, and those who have stayed have done so because *they* wanted to. So it is with every frontier. In places like Alaska, though, the numbers are smaller. That makes being an Alaskan more distinctive. Alaskans are proud and independent because they feel it just plain takes more guts to stick with a place like Alaska.

Imagine now what happened when one of the biggest oil strikes in history was made on the Alaskan North Slope. All but overnight, the whole world was very aware of Alaska. Before this, it had been a place of fascination and mystery. Its literary image outside was mainly in the boisterous poems of Robert Service and the dark fiction of Jack London. To the world at large, the state's problems were largely masked from view.

Since 1968, though, the entire world has known all about Alaska. We live in a world where money compels attention. The prospects of Alaskan riches have been the high points on the fever chart of its history. The financial impacts of North Slope oil were early and thunderous. The *Wall Street Journal* was on the story from the first known core samples out of Prudhoe Bay State No. 1. *Barron's* and the London *Economist* picked up the narrative. As the breadth of the Alaskan discovery grew, the *Journal's* closely read "Heard on the Street" column began a succession of reports on oil companies at the North Slope that would continue to appear every few months to this day—mirroring the market action from Alaska's new wealth, and touching off more of it.

The story quickly got bigger than that, and so did the amount of attention directed at Alaska. Soon, *Fortune* and others among the more reflective publications were moving beyond the purely business news to trace the impact of what was happening on the state itself. So were newspapers everywhere, including Alaska's own newspapers at first hand. And a broadening range of national magazines were looking at Alaskan society in general, its growing signs of culture shock—and the possible effects of the new oil boom on physical Alaska.

A searching debate over energy, economics, and the ecology had long since begun in the rest of the U.S. and in many other nations. Nor was environmentalism a new subject in Alaska. But nowhere, before 1968, had resource development and the environment come to such a sharp focus as an issue. Alaska now became a stage for a mighty confrontation over man's use of the earth. The drama would play for long years. The cast of characters—and the audience—would be worldwide.

4.
Why an Alaska Pipeline?

My son Stephen was a 17-year-old high school student when I began the research for this book. From a young age, he'd had a splendid ability—disconcerting, too, for his slower elders—to cut through to the heart of things and ask the One Important Question. We were in his backyard in California one time after I had just come down from the pipeline project. "Why," asked Steve, "do we have to build an Alaska pipeline anyway?"

"Oil is where you find it . . . and that ain't always where you want it." Some Texas wildcatter once said that, and he has been quoted often enough since. The wildcatters, independent oilmen, historically have been the front-runners in the search for oil. They first found much of what the large oil companies ultimately sold through their familiar branded gasoline pumps to us, the oil-consuming public. Col. Edwin Drake was a pioneer wildcatter who found the world's first commercial oil in Pennsylvania in 1859, long before anyone ever heard of John D. Rockefeller. Some seventy years later, when the major companies were well established in the U.S. and expanding around the world, Columbus M. "Dad" Joiner was the stubborn loner who discovered the East Texas oil field. It would stand for almost forty years more as the largest single American oil deposit.

In 1930 the United States still produced two thirds of the world's oil, and would produce more than half of it until after World War II. But long before that, the U.S. no longer *had* the largest known share of world oil. Beginning in 1908 with the British discoveries in Persia—now Iran—the balance of proven oil reserves had begun to shift overseas. By the late 1940s, Iran, Iraq, and Saudi Arabia together held more known reserves than the U.S. and Canada combined. Discoveries continued in other Middle East nations, in North Africa and down Africa's west coast, in Venezuela and the Far East, and in Russia. By the late 1960s, the U.S. had more than doubled its own proven reserves. But the rest of the world had found more than ten times as much. Total U.S. reserves were now only about 10 percent of the world total. Also, the U.S., with highly developed industries and a society accustomed to plentiful fuel supplies, was now consuming oil faster than it was finding it. The use of oil in the U.S. was far out of proportion to its population or its reserves, and was possible only because of sharply rising imports of oil from abroad.

Now the postwar geopolitics of oil came into play. The situation was simply that one half of the world produced most of the oil, and the other half used most of it. But into the 1970s, the major American and European oil companies still controlled most production around the world through concessions in oil-rich countries that had been developed with western funds and technology. The richest of these concessions by far were in the Middle East, where well over half of the world's total proven reserves were now located.

The producing countries had geography on their side. The oil, after all, was under their land. During and after World War II, the host countries and the major companies holding concessions had arrived at a standard fifty-fifty split of the profits. Before that, the companies had taken the major share. The host countries were now accumulating huge wealth. In 1960 the four major Middle East producing nations and Venezuela formed the Organization of Petroleum Exporting Countries—OPEC. Its aim was to control both supply and pricing of its members' oil. This was a radical jolt to a western world that had invented the oil industry and whose companies and governments had ruled all of the industry's market factors for a century. There was skepticism, though, that OPEC's strategy would work as long as oil's major markets were outside OPEC.

It worked. World oil in the 1960s became a chess game. Major companies, with their worldwide supplies centered in the Middle East, sought and found oil in non-OPEC countries. But as oil reserves increased throughout the Middle East, and in Africa, South America, and the Pacific, OPEC itself grew to a membership of thirteen nations. Some of the new members, like Libya and Indonesia, had ideas of their own. But whatever new agreements they gained individually from the oil companies generally became OPEC-wide before long. Many more companies were now operating internationally, and this broke whatever solidarity the original U.S. and European companies abroad had had in dealing with the producing countries.

The "energy crisis" is no brand-new fear. Warnings of coming oil shortages are as old as the modern oil industry. In fact, the world's oil reserves have seldom been more than twenty or so years ahead of its consumption. But into the 1960's, new and diversified supplies—and the tidy concession agreements—kept the threat at bay. Just as importantly, prices also stayed stable. Much like its food supplies, the world's supplies of oil and other forms of energy are sizable. Any "crisis" is likely to be one of availability and distribution rather than quantity. With diverse reserves of oil available with few serious interruptions over the years, where was the worry? In the U.S., though, it had been obvious for some time that oil discoveries and production were not keeping up with rising demand. More U.S. exploration and production, development of new energy sources, and conservation of use all would have helped. But little was done, by industry or through government policy, to combat the growing dependence on imports.

OPEC's strength today has shown how foolish this was. Rising nationalism in underdeveloped countries was adding to the threat. And the new presence of Israel in the midst of the Middle East brought everything to a climax. The western world created and supported Israel after World War II. Arab nations simply view Israel as an alien trespasser. The first Arab oil embargo against the U.S. and Europe, and the closing of the Suez Canal for nine years, came after Israel's victory in the 1967 six-day war. That embargo didn't stick because Iran and other non-Arab members of OPEC didn't join in it. Nor was the important question of oil prices really involved. But by 1973, a larger and more unified OPEC would use the next embargo by its Arab Members as a pricing lever of formidable effectiveness.

Half the globe away from the huge oil fields around the Persian Gulf, not many Alaskans ever expected their state to be a strategic piece on the chessboard of world oil. But a few did. And so did a few lonely evangelists in the world's major oil companies.

Oil oozing out of the ground in Pennsylvania in 1859 had created the oil industry. Eskimos in the Alaskan and Canadian Artic had been using oil seeps as a curious occasional fuel supply for centuries. Russian, British, and American explorers had seen traces of oil in the Arctic and on the Alaska Peninsula in the south as early as the first part of the nineteenth century. Alaska's first commercial oil well was drilled at Katalla on the Gulf of Alaska in 1901, and the small field's production was refined and sold locally for many years. Less successful attempts to find commercial oil were made all along the Shelikof Strait and the Gulf of Alaska until the 1930s.

The possibility of a North Slope oil field was known to the United States government almost from the time it bought Alaska. Several American companies tried to enter geologic survey claims on the Slope in 1921. Now the government was interested. Before World War I, it had been setting aside petroleum reserves to assure a fuel supply for U.S. naval ships: Elk Hills and Buena Vista in California in 1912, and Teapot Dome in Wyoming in 1915. President Warren G. Harding in 1923 added the largest area of all to the program: Naval Petroleum Reserve No. 4, a tract of 23 million acres on the western North Slope of Alaska. "Pet 4" was bigger than the state of Maine. The Navy surveyed Pet 4 in the 1920s and began drilling exploratory wells there during World War II. Its best estimates were that reserves might be as much as 100 million barrels—not really a commercial quantity in view of the remoteness and harsh working climate of the Slope.

Interestingly, though, World War II had also brought oil pipelining to Alaska. There had been pipelines there before—British Petroleum's David Henderson relates that a 48-inch waterline had been elevated on wooden piles near Fairbanks in the 1920s. The oil used in Alaska during the war had been shipped in from elsewhere. But small-diameter lines were built to move it around in support of Maj. Gen. Simon Buckner's Alaska Defense Command in its campaigns to push the Japanese back out of the Aleutians. One three-inch line went from the port at Haines up to Fairbanks, a distance of about 700 miles. A six-inch line traveled the short distance from Whittier to the military air at Anchorage. The Canol Pipeline was built 1,600 miles from Norman Wells to Whitehorse in Canada's Yukon Territory, then on to Skagway and over to Fairbanks. And the Alaska Scouts had done their northern pipeline survey for the Navy.

After the war, oil companies—especially those in the western United States—watched the Navy's continuing explorations on the Slope. But their attention was more on the Gulf of Alaska in the south. The interest at first was desultory. Even southern Alaska was an expensive place to look for oil. Finally one company began drilling a well on the Swanson River, which cuts across the Kenai Peninsula southwest of Anchorage. Like the rest of the Alaskan Gulf coast, the Kenai had been prospected early in the century. The company was one whose name would be heard again in Alaska: Richfield Oil Corporation of California.

Other companies became more interested. Leasing activity on the Kenai picked up. And in July 1957 Richfield's first well came in successfully—a highly rare event for a wildcat drilling program in new area. Further wells confirmed a field estimated at 175 million barrels of reserves. Natural gas deposits were also found and, a few years later, another oil field. In the early 1960s, several of the oil companies now active on the Kenai moved offshore into the Cook Inlet. The tide, wind, and icing conditions there were a fair preview of what was to be encountered soon in Europe's North Sea. In 1963, Shell Oil, operating for itself, Richfield, and Standard Oil of California, brought in the first successful well in the inlet. Other finds followed.

By the late 1960s, more than fifty onshore wells were operating on the Kenai Peninsula. More than fifty more wells, drilled from a dozen platforms, were producing in the Cook Inlet. From a standing start, Alaska now ranked eighth among U.S. oil-producing states. It was still, though, a minor oil state. The Kenai and Cook Inlet reserves together were something over 2 billion barrels—barely 5 percent of total U.S. reserves at the time. Production was a slightly lower percentage of total U.S. volume. Oil in the south of Alaska hadn't been that hard to find. But there didn't seem to be all that much of it. Producing conditions were costly and tough. And markets were far away for the volume produced. Nor had the full importance of OPEC's post-1967 strategy really sunk in around the world.

Meanwhile, British Petroleum and the others had come and gone on the North Slope. In early 1966, however, Richfield Oil merged with Atlantic Refining Company. ARCO now had Richfield's North Slope leases. ARCO geologists looked back over the early geologic work at the Slope. They thought they saw something the others had missed. But they couldn't convince the ARCO and Humble managements to go ahead with a drilling program. "They kept pushing, pushing, pushing to drill," remembered Hal Peyton. "It almost cost them their jobs. Well, not their jobs . . . but they had a loss of credibility within the company to the point that they were losing effectiveness."

A North Slope program was begun in 1966. The first well, the Susie Unit No. 1, was a dry hole. There would be one more try. The rig was moved sixty miles north and a new well was put down. It was Prudhoe Bay State No. 1.

When the Prudhoe Bay field was confirmed at up to 10 billion barrels of reserves in mid-1968, it shot Alaska not only into the headlines and stockmarket tables of the world—but also, almost overnight, into a pivotal position in the politics of world oil. Completion of the Alaska pipeline this year is only the first step. Alaskan reserves at Prudhoe Bay alone are five times the state's total proven oil reserves before 1968. Alaska will soon be producing more oil than any other U.S. state except Texas. At the level that U.S. production rates are predicted to reach in 1980—perhaps 10 million barrels per day—Alaska could be supply as much as 20 percent of the country's domestic oil.

But that may be just the beginning of Alaska as an oil source. How much of a beginning? That is not easy to say. The late 1970s are a critical time for U.S. oil exploration. You must match varying estimates of future U.S. discoveries against estimates, which also vary, of Alaska's potential oil and gas. Trying to reckon oil reserves accurately is like trying to count snowflakes in a blizzard. The trend, since the 1968 discovery, has been toward ever-higher estimates in Alaska. In contrast, over the same period, estimates of total U.S. reserves and potential discoveries have been sharply cut.

As recently as 1974, for example, the U.S. Geological Survey had calculated undiscovered—but probably recoverable—U.S. oil and natural gas liquids at a range of 200 billion to 400 billion barrels. Undiscovered but recoverable natural gas was put at a range of 990 trillion to 2,000 trillion cubic feet. But just a year later, the U.S. geologists abruptly lowered these calculations. Potential oil and gas liquids were now put at a range of 61 billion to 149 billion barrels. Natural gas was put at 332 trillion to 655 trillion cubic feet. The National Academy of Sciences, at the same time, made similar estimates: 113 billion barrels of oil and 530 trillion cubic feet of gas.

Industry figures were in a like range. An Exxon USA study, in March of 1976, put the figures for oil and gas liquids at 118 billion barrels and natural gas at 582 trillion cubic feet. These figures almost matched the amounts of oil and gas—122 million barrels and 477 trillion cubic feet—that had been produced over the U.S. oil industry's entire history up to 1975. Also, said Exxon's vice president for exploration, J.D. Langston, it would take improved technology and higher price structures to reach those levels of development. At present, the practical limits were really about 63 million barrels of oil and 287 trillion cubic feet of natural gas.

Remember, too, that these are estimates of oil and gas that have not yet actually been found. *Proven* U.S. reserves are pegged much lower: about 35 billion barrels of oil and gas liquids and 220 trillion cubic feet of natural gas. At current rates of U.S. production and consumption, that's only enough to last about five or ten years. Without imports, U.S. oil and gas would be used up in less time than that.

But now, with North Slope oil and gas, we can look at these figures from the opposite direction. Until just recently, Prudhoe Bay's oil production has not been included in many of the estimates. Until this year, none of it has actually been produced. There was no way to transport it to market. Now there is an oil pipeline. Prudhoe Bay, in studies following up the initial DeGolyer & McNaughton report in 1968, is presently rated at 9.6 billion barrels. So the Prudhoe Bay field alone already amounts to nearly 30 percent of all U.S. oil reserves. More optimistic estimates are that the field may contain as much as 15 billion barrels of oil. So its contribution could be much larger. One reason for this is that estimates of proven reserves at Prudhoe Bay so far have been based on just one oil formation, or pool, in the field. But that is only one of at least three oil and gas formations known to be in the field. Until all of the formations are fully explored, the true value of Prudhoe Bay can't be known for certain.

Why an Alaska Pipeline?

Prudhoe Bay, moreover, is only one small part of the North Slope. Eastward is the Arctic National Wildlife Range, which for environmental reasons may never be explored. Westward lies the Navy's Pet 4. Despite earlier pessimism, new estimates after the successes at Prudhoe Bay are that Pet 4 may contain up to 10 billion barrels of oil and 13 trillion cubic feet of natural gas. Those were moderate estimates. Wilder guesses range much higher. But the fact is that barely 5 percent of the total area of the North Slope has been at all heavily explored for oil so far.

There's more. The North Slope, as big as it is, is only one of fifteen known sedimentary basins under Alaska and its offshore waters. Northeast and northwest of the Slope, in the Beaufort Sea and the Chukchi Sea, are two other large sedimentary basins. Down the seacoast of the Bering Sea—in Kotzebue Sound, Norton Sound, Kuskokwim Sound, and Bristol Bay, and extending inland from these areas—are other basins. Along the south Alaskan coast—the Alaska Peninsula, the lower Cook Inlet, and in the vast Gulf of Alaska—are yet more basins. And several of Alaska's interior regions have shown evidences of oil and gas formations. One estimate of Bering Sea oil is 27 billion barrels. The Gulf of Alaska has been pegged at up to 20 billion barrels.

Such figures, of course, can only be seen now as highly speculative. But the potential is nonetheless dizzying. All told, even the more moderate estimates of total undiscovered oil reserves in Alaska range up to 80 billion barrels. If true, Alaska alone could double and possibly triple total U.S. oil reserves. The same numbers game can be played with Alaskan natural gas. The known reserves of natural gas at Prudhoe Bay now are 26 trillion cubic feet, or almost 12 percent of all known U.S. gas reserves today. At some point this year, a decision will be made on another massive pipeline system to bring that gas out. Beyond this, total undiscovered natural gas reserves in Alaska have been calculated as high as 450 trillion to 1,000 trillion cubic feet. Again, if realized, this would match or double all estimated undiscovered natural gas reserves in the U.S. today.

Is Alaska shaping up as America's own Persian Gulf? Not nearly, sad to say. If the most flamboyant hopes for actual reserves of oil and gas in Alaska came true, total U.S. reserves would still reach a level of only about half of the combined proven reserves, as known today, of all of the Persian Gulf nations. So it would take major discoveries in other parts of the U.S., perhaps offshore in the

Atlantic or Pacific, to put the U.S. anywhere near par in total oil supplies with the Middle East. Some contradictory trends are also at work here. The Middle East's known oil reserves themselves may or may not have peaked out—there is still some unexplored territory in that part of the world, too. In any case, most exploration and production in the temperate Middle East has come easily and cheaply. Neither seeking oil nor bringing it to market comes cheap in the major areas where the U.S. must now do it. This is especially so on our frontiers. The Arctic tundra reaches of the North Slope and the tempestuous waters of the Gulf of Alaska are just two tough examples.

Why, then, do it? And why consider doing more of it? Perhaps the best answer is: insurance.

The U.S. now depends on imported oil for more than 40 percent of its total needs. Most of that amount comes from the OPEC nations, mostly the Middle East and Venezuela. By 1980, it's been predicted by analysts like Charles T. Maxwell of Cyrus J. Lawrence, Inc., both imports and the OPEC portion of them will be above the 50 percent level. That date is scarcely two years away. Only immediate or extremely short-term solutions can counteract it.

The Alaska pipeline, as it goes into operation this year, is the first of those solutions. At a throughput of 1.2 million barrels a day, for instance, it will just match our 1976 imports from Saudi Arabia. Maxwell cautions that North Slope oil production initially will only meet the current increase in U.S. demand for oil, which is already rising at an annual rate of 500,000 barrels a day. It will only satisfy a small part of the base demand itself. Conservation of use might slow this trend. But Maxwell says that the real importance of Alaskan oil lies in future discoveries, which could substantially lessen the need for imports.

The pipeline through Alaska is a "short-term" solution that has taken nearly a decade to put into effect. It was proposed early in 1969. It might have been completed by the time of the 1973 oil embargo. How this might have dampened OPEC's supply and pricing power than, we'll never know. We do know that the abrupt fourfold rise in world oil prices in early 1974 was a major factor in putting the U.S. and a large part of the rest of the world into a long-term economic recession and inflationary cycle. So the availability of Alaskan oil can be seen as a U.S. defense against future embargoes and their results—an insurance policy in a world where resources like food and fuel are used as political and economic weapons.

Yes, there were other solutions. There still are. There are other forms of energy. Some could be available fairly quickly. Others might take decades—even the development of proven oil and gas can take five or ten years. Alternatives to oil and gas have been known for a long time. Little is being done, even today, to bring them along. The government blames industry. Industry blames the government. Yet after ten years of warnings from OPEC, the United States is only now taking some steps toward a coordinated energy policy.

So the development of a major oil source like Alaska, and the building of a pipeline to carry it, becomes necessary.

Steve is a sophomore now in environmental studies at the University of California at Berkeley. There may be a better answer to his question. But his generation will probably have to find it.

5.
1970...Shutdown

As 1970 began, the ice bridge across the Yukon River was in place. Hopeful of gaining its permits any time now, the Trans Alaska Pipeline System was issuing letters of intent to the contractors who would build the construction haul road. Bids were also coming in from construction companies to build the pipeline itself, in hundred-mile sections both north and south of the Yukon. TAPS held off from letting these contracts yet. But the heavy equipment and the materials for the construction camps along the haul road route had to move north before winter ended.

The crews were using portions of the old Hickel Highway where they could and were pushing new winter trails through the chill landscape where they could not. Jane Pender of the Fairbanks *News-Miner* went up to Bettles Lodge on the John River in February. Road and pipeline permits may not have been issued, she wrote, "but hundreds of men and 400 tons of equipment a day have been pouring up the ice road since January." Without permits in hand, TAPS was taking a calculated risk. But it had to. If the massive machinery and the camp units and supplies weren't moved over the frozen ground by the spring thaw, there was no way to get it all up there until freeze-up the next fall, and the roadwork could not begin.

Morris J. Turner, an official for the Department of Interior, also watched as the men and machines headed across the Yukon that winter. He wrote a detailed report: "Mobilization of the TAPS Haul Road." His firsthand observations were instructive when viewed against later environmentalist outcries about the ravages the early pipeline work had committed on Alaskan terrain. Said Turner: "I have been impressed with the mobilization effort on the part of TAPS and its contractors to date. In most areas very little damage has been done to the environment and we expect almost no degradation because of the care that has been taken. I consider the camps as models. Should a permit for the road be allowed and the same care exercised in the ensuing construction, a milepost will be established in cold region engineering."

It would be some time before his prediction could be tested. The environmentalists, and the natives of northern Alaska, were to have the last word—or at least the next word.

On January 1 the National Environmental Policy Act of 1969 went into effect. NEPA had been making its slow way through Congress for most of the past year. Few people within TAPS had taken any special note of the legislation's converging track with their efforts to get permits for an Alaska pipeline. "We were right in the middle of the application process at the time the act was passed," says one attorney. "But nobody knew what the hell that law meant." The intent of NEPA, in fact, was to provide safeguards over any project that crossed federal lands. A key provision of NEPA was the requirement of an environmental impact statement on any such project, plus evidence that the alternatives to the project had been reviewed, and a summary of those alternatives.

The previous September, Alaskan natives with land claims across the proposed route of the pipeline in central Alaska had signed waivers to let the project go through their property. Now they complained that an assurance of jobs on the project for their people was not being honored. On February 4 two native villages in the Alaskan interior sued TAPS in Alaska's Superior Court and withdrew their waivers. On March 9 five villages filed a second suit in federal district court in Washington, seeking to enjoin Interior Secretary Hickel from granting road and pipeline permits over the claimed land.

Next it was the environmentalists' turn. On March 26 the Wilderness Society, the Friends of the Earth, and the Environmental Defense Fund also filed suit in Washington's federal district court. They charged that TAPS was requesting an excessive right-of-way width in violation of the Mineral Leasing Act of 1920, which restricted rights-of-way to 50 feet—25 feet on each side of whatever was being built. They also sought to consolidate their suit with the first of the native suits. On April 1 District Court Judge George L. Hart issued a preliminary injunction preventing the pipeline from crossing land at Stevens Village'on the north bank of the Yukon. On April 5 three of the native villages filed a third suit, this time against the state of Alaska. Now the meaning of NEPA was about to become clear. On April 6 the environmental groups amended their suit to charge that the Interior Department had not complied with the impact statement requirements of the new National Environmental Policy Act. On April 13 Judge Hart issued a second injunction, against the haul road.

In just two months of legal maneuvers, both the Alaska pipeline and the road needed to build it had been stopped dead in their tracks.

The state of Alaska tried to save the situation. Keith H. Miller as lieutenant governor had succeeded to the governorship when Walter Hickel went to Washington. Miller came up with an 1866 federal law that allowed states to establish a right-of-way over unused lands in the public domain. The state could authorize the haul road itself. Miller's scheme was that the state would hire TAPS as its contractor to build the road, and TAPS would later reimburse the state for its costs. Alaskans cheered—or at least those favoring the pipeline did. "The Governor Was Right," headlined the pro-development *Anchorage Times*. In Fairbanks the *News-Miner* praised Miller's action in "cutting through the red tape."

On the other hand, reported the *Anchorage Times*, TAPS officials themselves were "stunned" by Miller's proposal. So was the Interior Department. It sought a Justice Department opinion on the legality of what Alaska sought to do. Justice said the action was legal. Interior nonetheless urged Governor Miller to hold off. TAPS settled the matter by declining to go along with the state. Legal or not, Miller's plan would not have cleared away the federal injunctions against either the road or the pipeline. As a practical matter, the road was useless to TAPS without a pipeline permit, and TAPS would have had to pay for it even if the pipeline were never built.

In mid-June, with the project hopelessly stalled, TAPS began terminating some of the letters of intent to the road contractors. The machinery was stranded north of the Yukon, and so was the string of construction camps that had been built. Financially, for the builders who had committed their own funds, it was a bad situation. "Finally," relates Clyde Johnson, "this resulted in us buying the equipment and the camps from some contractors . . . it would have busted some of them otherwise." The camps with their airstrips were used for TAPS' continuing surveys and soils explorations. The equipment was "mothballed"—weatherized and stored away to wait out the delay.

"No one realized then," Johnson adds, "that it would be another four years before that equipment could be used."

In the midst of all of this, technical arguments over the pipeline's design were developing. After the hearings in Fairbanks the previous summer, the Interior Department had appointed a study group headed by Dr. William J. Pecora, director of the U.S. Geological Survey, with members including engineers from the TAPS staff and scientists from USGS and other Interior agencies. Their first task was to organize a further core drilling program in the Copper River Basin, an area more than halfway down to Valdez from Fairbanks, and several hundred miles south of the Arctic Circle, that Pecora felt would present some especially troublesome soil instability problems. In December the study group had set up a formal base of operations in Menlo Park, California, and met there to analyze the new Copper River data.

The thorniest technical issue was the burying of a pipeline in permafrost. At this early point in its design, TAPS still planned to bury most of the line. True, its own soil borings in 1968 and 1969 had shown that permafrost might be found along as much as 85 percent of the proposed route. But the research was also showing that "permafrost" was a broad term for a ground condition with many variations. The classic definition of permafrost as "perennially frozen ground" contained many traps. Permafrost may lie in regions that never come close to thawing, and so may be extremely stable for construction. Some permafrost, too, is almost solid rock or soil with almost no ice content, and these areas will stay stable. Other permafrost has a high ice content, and drains poorly, so it is unstable when thawed.

TAPS felt that burial was possible in all but the most unstable permafrost areas. By late 1969 this design concept had been developed to the point that all below-ground pipeline was to be set in gravel insulating pads, buried at a depth of six to eight feet. River crossings, including the Yukon, were to be buried also, deeply trenched below the scour level of the stream beds. TAPS still had no evidence that no more than 10 percent of the line would have to be elevated to avoid unworkable permafrost or other ground problems. With the soil surveys still going on, it wasn't sure yet where the raised sections would have to be. But it had been designing to meet them. One concept was to lay the line over a gravel berm several feet thick and insulated between its base and the ground surface below. Another idea was to raise sections of the line entirely off the ground on wooden pilings, and insulate the pipeline itself with plastic foam to hold its heat in.

Scientists disagreed about burying the line. In a USGS publication, Interior's Arthur H. Lackenbruch warned about the "thaw plug" around a hot oil pipeline in permafrost—the cylindrical mass of earth that would be heated up by the pipeline passing through it. In a few years, said Lackenbruch, the "bulb" or outer diameter of this mass would extend out twenty or thirty feet, and would keep expanding, especially in Alaska's warmer south. The pipeline, if suspended now only in sludge, would sag to its breaking point. Insulation, or even cooling the oil, would make little difference—snow on top of the line would melt, and nearby plant life would be affected. Science writers picked up the report and ran with it. Wrote Warren Kornberg in a syndicated Science Services article: "What would be left along the oil line will be a muddy, slush-filled, lifeless scar running 800 miles across the face of Alaska."

Pecora, meanwhile, had gone to Alaska to tour the proposed route and talk with Max Brewer at the Naval Arctic Research Laboratory. Brewer was shortly to join the Menlo Park study group at the suggestion of Dr. William R. Wood, president of the University of Alaska. Pecora and Brewer also concluded that not enough research had been done on burial techniques in permafrost to assure safety. Brewer felt that none of the pipeline could be buried safely north of the Brooks Range, including the gravel beds of the Sagavanirktok River. It was Pecora's opinion that as much as 25 percent of the line would have to be elevated, and the USGS said later that the pipeline should not be buried in *any* permafrost area. George Gryc, chief of the USGS Alaska Geology Branch, acknowledged the pipeliners' view that a buried pipeline would be cheaper to build, and that the integrity of an underground line is normally more secure. But, Gryc said, TAPS hadn't yet proved that this would work in Alaska.

As to raising the line, the USGS staff also questioned if permafrost could even support wooden pilings driven into it. The permafrost, as it contracted and expanded, might reject the pilings, heaving them back out of the ground. In fact, as 1970 went on, TAPS itself was having doubts about the simple wooden pilings that had been a usual way to support raised pipelines in northern climates. The TAPS engineers and their consultants were looking at more advanced ideas. Esso Production Research, a unit of Humble Oil in Houston, was doing studies. And William Black at Woodward-Lundgren was directing research into steel pilings that used air convection and heat transfer techniques to maintain below-ground temperatures.

Engineering refinements like this were a normal part of any project's design. But it was increasingly clear that the Alaska pipeline was being subjected to more advanced scrutiny than any pipeline ever built anywhere. The Interior Department had issued a revised set of construction stipulations that now ran to 60 pages. These guidelines were technical, environmental, and even social. They specified the care to be taken with wildlife habitats and stream beds. Pump stations and other permanent installations were to be built to harmonize with their natural settings. The route was to be surveyed ahead of construction for evidence of Alaska's archeological history, and such sites were to be either preserved or dug out and their artifacts taken safely elsewhere. Revegetation, of course, was required over the pipeline route and at gravel pits and other material-gathering sites. The project was to follow equal opportunity employment practices, particularly for Alaskan residents and most particularly for Alaskan natives.

Earthquake precautions were to be extensive. The line was to be built to withstand earthquakes and instrumented to detect seismic shocks as they occurred. TAPS had already anticipated this. Sections of the pipe from Japan had been shipped to the Structural Research Laboratory at the University of California at Berkeley for welding and bending tests directed by the lab's Dr. Graham Powell. It was proven the pipe could be welded effectively at temperatures as low as minus 35°—"not a major problem," noted one technician, "with the exception of discomfort to the workers."

The bending tests were torturous. The largest known testing machine at that time had been built for 36-inch pipe, so a new machine was built. It wrestled the pipe sections in every way that an earthquake or sagging permafrost might possibly do it. Pipe was compressed axially—from end to end—at forces up to 2.55 million pounds, and laterally—from the sides—to 750,000 pounds. Varying temperature differentials were applied simultaneously to the inside and outside walls of the pipe, and a range of internal pressures also. The object of these tests was to see how much stress the pipe in a fixed position could take before it first wrinkled and then cracked. A 31.5-foot section of pipe proved able to take a deflection, or degree of bend at its midsection, of 1.2 inches before it wrinkled. At this point, axial force and internal pressure were increased. The deflection reached 33 inches before the pipe actually cracked.

Wrinkled pipe was also tested hydrostatically, by forcing water into it under pressure. It reached 95 percent of its minimum yield strength without leaking, and this was 32 percent more pressure than the pipe would be subjected to when the pipeline was operating. Lengths of pipe were tested to 125 percent of yield strength before they gave way. The testing at Berkeley was designed to put stress both on the girth welds around the circumference of the pipe sections as they would be joined during construction, and on the longitudinal welds made when steel plate was rolled into pipe at the Japanese mills. Later, after the pipeline was laid in Alaska, each section of several thousand feet would be blocked off and hydrostatically tested again to assure the integrity of its welds.

No pipe for an oil line had been put through such rigorous testing before it was delivered to a project site. Usually, the pipe would have been factory-tested by its manufacturer, and a pipeline builder would then confirm that it met the specifications he had ordered according to standard sets of industry guides. But the Berkeley pipe tests were of a piece with the way the whole project was unfolding. Permits had been a fairly routine matter in the past. Construction methods had become standard. Revegetation of a pipeline route had long since been an industry practice—and, in temperature climates, what the builders did not put back into place, nature did. Detailed building stipulations in advance of a project had never been issued before. Nor had there been such thorough review of a design still under development. All of these traditional procedures of pipelining were changing sharply in the first years of the Alaska project.

The pipeline company itself was facing change. TAPS had grown unwieldy. Joint ventures were a common approach to projects like pipelines—in some cases, they still are. But no pipeline had required so much administrative, legal, and technical work before construction started, nor the support of such a roster of outside consultants. The original three-company venture had worked well enough. Humble Oil, at the start, had even proposed that it would design and build the project as manager for the other companies, but ARCO and British Petroleum had preferred to keep equal management roles. Then TAPS had expanded to an eight-company venture, operated by a far-flung committee of representatives from each owner company. Operating decisions, especially in contracting and spending matters, became awkward.

And another corporation—The Standard Oil Company (Ohio)—was joining the project. The North Slope was still reshaping the oil industry. In late 1969 Sinclair Oil Corporation had been merged into Atlantic Richfield. To settle a Justice Department complaint that the merger would give ARCO too much weight in some U.S. markets, Sinclair's properties in the eastern U.S. were sold to British Petroleum. In January 1970, BP in turn made a deal with SOHIO, transferring all of its North Slope leases, plus its new American refining and marketing operations gained from Sinclair, into SOHIO in return for a 25 percent interest in SOHIO's stock that would increase to 54 percent as Prudhoe Bay production increased.

It was now decided to incorporate the pipeline venture. In August 1970 Alyeska Pipeline Service Company was formed as a nonprofit contractor to design and build the Alaska pipeline—and, ultimately, to operate the completed system. Home Oil of Canada left the project. The ownership percentages shifted again. The companies, or their pipeline units, now held shares like this: SOHIO, 28.08 percent; ARCO, 28.08 percent; Humble, 25.52 percent; Mobil, 8.68 percent; Union Oil, 3.32 percent; Phillips Petroleum, 3.32 percent; and Amerada Hess, 3 percent. The British Petroleum interest for now was included in the SOHIO share. Alyeska kept its engineering base in Houston, and its Alaska office under David Henderson. But it established an Alyeska corporate headquarters in Bellevue, Washington, near Seattle. and it began setting up its corporate management.

Edward L. Patton now arrived on the Alaska pipeline project as Alyeska's president and chairman of its construction committee. Patton's career specialty had been the building of big projects. A Virginian, he had joined Standard Oil of New Jersey in 1938. After several U.S. assignments, he supervised the construction and operation of a large refinery and terminal complex for ESSO Norway. Since 1966, Patton had been in California, managing the building and start-up of Humble Oil's newest U.S. refinery at Benicia, north of San Francisco.

Ed Patton's style is to be blunt and candid—to tell it like it is, in public or wherever. The pipeline was completely stalled, and the first task of the new Alyeska's management was to get it moving again. Patton would all but live in Washington, D.C., from now on in the effort to qualify the project in the eyes of its critics in government and elsewhere, clear away the lawsuits, and get the permits to start building. There was one matter in particular that had to come first. In September 1970, speaking to the Anchorage Chamber of Commerce, Patton cut through the maze of issues delaying the pipeline. The permit applications could make only limited progress, he said, until the Alaskan native land claims were settled.

6.
1971...Political Paper Work and Native Claims

The Department of the Interior is found between C and E Streets in Washington, D.C. It watches over all of the fifty states, and its cavernous building seems big enough to contain several of them. For the Alaska buff, DOI is a trove of documentary treasures. On a basement level of Interior's library, ranging across shelves and spilling over onto handcarts and the floor, the adventurer can set off upon the long and voluminous paper trail of the Alaska pipeline.

Clearly, 1971 was a vintage year for pipeline paper work. Here, dated in January, is the first draft of Interior's environmental impact statement for the project, as required by Section 102(2)c of the National Environmental Policy Act of 1969. It runs 196 pages, plus the 60 pages of revised construction stipulations. Here, with February and March dates, are the testimony and exhibits from hearings held on the impact statement. Testimony scheduled for two days in Anchorage had taken five days to complete. The scheduled two days in Washington had also run overtime. The total record of the hearings is some 12,000 pages long. The Anchorage sessions alone produced three volumes of supporting documents in seventy-one separate exhibits.

69

Testimony by the project's dozens of sponsors, opponents, and otherwise interested observers covered the whole by-now familiar range of pipeline issues. The countless words that pour forth in government hearings seldom join the literature of the ages. But one statement at the Anchorage hearings achieved a certain sort of immortality. Geoff Larminie, the far-roving Irish geologist for British Petroleum, was now the company's area manager in Alaska. Larminie's travels had shown him firsthand that development and the environment can coexist in harmony, and he was a fervently articulate champion of the view that Alaska was not going to crumble under the weight of progress. "Indeed it all too often seems that many of these critics who so freely use the term 'ecology' know as little about ecology as they know about Alaska," Larminie told his Anchorage listeners.

"The true expert," Larminie went on, quoting Edmund Burke, " '. . . provided he keep his mind open to new ideas . . . is fully qualified to make responsible decisions without accounting for his every move to the inexperienced and ill-informed outsider.' In other words, why should we have to continually account for our knowledge, and they never account for their ignorance?" Larminie left Alaska shortly for further BP assignments around the world. His words stayed behind him, engraved on a plaque in prominent view at the Anchorage Petroleum Club.

Hearings like this, and the tens of thousands of pages of documents that punctuated them, would go on for more than another two years before the project could start. The documentation would continue long after that. Thomas C. Grimes came to Alyeska from General Electric as manager of information services in 1972, and took over the project's technical library that had been begun by Chris Whorton. "One of the things Grimes has done is calculate the size of our submissions to government," said Hal Peyton. "It's a stack of paper fifteen *miles* high. That's even beyond my comprehension, and I've had to review most of it."

Interior's 1971 draft impact statement had been prepared largely during the lively reign of Walter Hickel as secretary. But Hickel was no longer around to announce this weighty analysis of his own home state. In fact, he was back in Alaska. Despite the doubts over his environmental qualifications a year earlier, Hickel had so enthusiastically embraced ecology and other counterculture causes of the day that an outraged President Nixon had fired him out of the government this past November. Hickel, though, for all that he had moved cautiously, was mainly regarded a pro-pipeline at the end. His successor at Interior, reliable Republican stalwart Rogers C. B. Morton, was harder to figure about the pipeline. Morton was planning a trip to Alaska in June.

The draft impact statement was a good traveler's guide for him to take along. In quasi-technical, environmental terms, it told much of what Alaska was all about in the first place. It dealt with the permafrost hazards and the possible impacts of construction on the land, wildlife, and human society of Alaska. It outlined the proposed pipeline and described and evaluated the alternatives to it. The *Manhattan* voyages were summarized. Possible Canadian overland routes were discussed. The overall preliminary conclusion was that an Alaska pipeline could be built with containable impact on the state's environment.

While the pipeline studies and the *Manhattan* tests were going on, there had been no shortage of other schemes to move Alaskan oil being offered from outside the industry. They ranged from the practical to the sincere to the bizarre.

One of the more practical alternatives to a pipeline or ice-breaking tankers had been proposed to individual oil company members of TAPS by General Dynamics Corporation. This was a fleet of 170,000-ton nuclear submarine tankers that could move at will, it was claimed, under the ice of the Arctic Ocean to a port in Greenland. Loading problems, cost, and uncertain navigability were the drawbacks, and Greenland was hardly closer to major markets than Alaska. There was another nuclear scheme that really raised ecological hackles. Dr. Edward Teller, the physicist, suggested clearing out a new deepwater harbor at the North Slope—the quick way, with nuclear explosions.

Extending the Alaska Railroad to the North Slope and moving the oil in tank cars was also suggested. Volume was a problem. The Interior Department calculated that this would require sixty-three trains per day in each direction, with one hundred cars in each train. This would be an environmental burden at best. Its economics were also poor, since the sixty-three northward-bound trains each day would all be "dead-heading"—traveling empty. Another suggestion was a monorail to move the oil itself, or at least to haul the pipeline's building materials. It was rejected for similar reasons. What about trucks, then? Interior figured a truck fleet of sixty thousand vehicles would be needed, and an eight-lane highway through the wilderness to carry them. Again, environmental factors and the economics of one-way cargoes overruled this scheme. Also, there were scarcely that many oil trucks in operation all over the United States.

Perhaps North Slope oil could take to the air. Both the Boeing Company and Lockheed Aircraft Corporation studied the use of jumbo jets then under development. But Alaska's winter air conditions could interrupt schedules. And the Alaska airways would be filled with about thirty times the air freight traffic that covered all the rest of the United States. Less seriously, oil-carrying blimps were suggested. Again, these were hardly an Arctic aircraft.

Many schemes to improve the pipeline itself were suggested. The oil, said one letter writer, should be put in aluminum-foil containers and floated through the line. Others said it could be sent through in air-propelled canisters the way money used to be shot around in an old-time department store. An Alaska newspaper report in 1969 said the pipeline could be used to develop two resources at once: bringing tubes of copper ore, floated in the oil, from the bornite deposits at Shungnak in the interior of Alaska.

Another suggestion was to build a pipeline big enough to hold tractor trains carrying North Slope oil in drums. One drawback: finding crews willing to make the trip from Prudhoe Bay to Valdez through, in effect, an 800-mile subway tube. There were also schemes to insulate the pipe from the permafrost, including one that would have challenged its designers: a triple-walled pipeline with the oil line itself resting in a vacuum. It was also suggested, and a lawsuit had to be dismissed before its sponsor gave up, that the best environmental protection for the pipeline would be to string the whole thing from 50-foot towers like a high-tension electrical system.

Leak detection also came in for much attention. Alaska's Senator Gravel in 1969 proposed using a heat detection satellite already in polar orbit over Alaska to sense leaking oil. In fact, satellites do form one of the backup communications systems over today's completed pipeline. But the most imaginative detection idea came also in 1969, when a bird trainer offered to train sea gulls to fly over the pipeline route looking for leaks. It wasn't made quite clear how the birds would then report any leaks that they found.

Interior's final impact statement was still another year away.

Meanwhile, in July, Alyeska's first full-dress project description filled in more of the pipeline's engineering details. Its three main volumes and another twenty-six volumes of appendices also filled another shelf at Interior's library. But the continuing engineering work on the project was evident, and so was the design progression that was resulting from it. Soil borings had now been made or were being made roughly every third of a mile along the route. More and more difficult permafrost and other terrain problems were being found. And so portions of the pipeline were rising above the ground from one end of the project to the other.

Alyeska now planned, according to its 1971 description, to elevate almost 20 percent of the line—about 136 miles. But, significantly, its maps also showed another 30 percent of the route where decisions to bury or elevate still had to wait for more study data. The research on air convection and heat-transfer systems in above-ground steel pilings was not yet complete. But the design now included a system using underground thermal piles with above-ground condensers to keep the soil chilled around some sections of the buried pipe.

Some of the buried river crossings were also getting a closer look. There was now a new option for the Yukon River. Alyeska still believed it could bury the line in Rampart Canyon deeply enough to avoid the Yukon's heavy scouring at spring breakup. But the state of Alaska now wanted to build a highway bridge across the river. If Alyeska would help pay for the bridge, the pipeline could be strapped to one side of it. The costs would probably be less than digging under the Yukon. The decision would come soon: to use the bridge.

Alyeska, at the Interior Department's request, spent the rest of 1971 amending and adding to its project description. Interior, to conform with NEPA, worked on its final impact statement. The project's delays were becoming habitual. And the state of Alaska, because of that, was in financial trouble.

The $900 million Alaska received in the 1969 lease sale was like a sudden inheritance falling on a young person who'd never known money before. Alaska went on a spending spree. The state budget was almost immediately doubed to more than $300 million for the next fiscal year. It kept rising from there. Good intentions were rampant. Elaborate state planning studies were commissioned from prestigious groups: Stanford Research Institute and the Brookings Institution. Wall Street was lusting after Alaska's new money, and long-range investment programs were discussed. But mainly the money poured out in a jumble of disconnected splurges whereby Alaska would catch up on the services and facilities that states in the Lower 48 already enjoyed. After all, the pipeline would be finished in three years, and state finances would then be assured long into the future as the income poured in from North Slope production royalties and pipeline taxes.

By the time the extent of the delay sunk in, Alaska was almost broke. It had legislated a hike in the production royalty from 4 percent to 12 percent, more in line with the take in other oil states. But this meant nothing until the wells flowed, which had to await a pipeline in which to put the oil. William A. Egan, a Democrat who had been Alaska's first governor after statehood, was returned to Juneau in late 1970 after the four-year term split between Republicans Walter Hickel and Keith Miller. Egan had a new solution. He wanted Alaska to take over the pipeline, financing the transfer with a $1.5-billion bond issue. Alyeska would still build it, but the state would own it. This might not get the project built any faster. But Alaska would now have a "bankable" asset to pledge as collateral for loans to meet its current urgent operating needs. It would also control its own oil resources from wellhead to shipping terminal, which the state officials saw as another financial plus.

Alyeska met with the governor in November. Gently the pipeline's owner companies pointed out the holes in Egan's scheme. The state had no assets with which to buy the project. It could only finance the purchase of the pipeline if the oil companies themselves guaranteed the needed bonds. In fairness to their own shareholders, not to mention the legalities of the idea, the companies could not do that. Nor were there any precedents for such an arrangement. The Alyeska companies gave Egan all the assurances they could that, once the lawsuits had been satisfied, both the building of the pipeline and its operation would be carried out with as much financial and social benefit to Alaska and its people as possible.

Besides, Alyeska told the governor, the pipeline was no longer a billion-and-a-half-dollar project. More than two years had passed since the last cost estimate. Research and engineering costs, some of which could not have been anticipated at the time of the early estimates, were accumulating. Inflation was hitting at every part of the project—materials, equipment, labor. There was faint comfort in the $100 million in pipe that had already been delivered, which looked cheaper every year. But Ed Patton warned that the estimates could only be tentative until the project actually got under way. In any case, the new estimate had risen to $3.5 billion. All other arguments aside, the pipeline by now was pretty clearly priced out of Alaska's reach—even if the state were as rich as Texas or one of the Middle East sheikdoms.

There were also some new legal hurdles. Another suit based on NEPA had been filed earlier in the year by the Cordova District Fisheries Union, which feared damage to its fishing grounds by the tankers passing in and out of Valdez. The proposed tanker traffic was also bringing Canadian environmentalists into the fight. Judge Hart had ruled that they could not do this, but another court overturned the judgement, and a Canadian suit was filed requiring the Interior Department to consider possible environmental damage to Canada's west coast as tankers passed between Alaska and California. Judge Hart now allowed both Alyeska and the state of Alaska to intervene and take an active role in the suits against them by the environmental groups and the Cordova fishermen.

As Ed Patton had said, though, none of the paper work and court battles would mean a damned thing until the native claims were resolved. Alaska's centuries of social history had come to an unlikely focus on, of all things, an industrial enterprise. And that enterprise couldn't go forward until this ancient grievance was met.

American federal policy toward Alaska in its long ninty-year struggle from Russian colony through district and territory to statehood was best described as one of offhand neglect. In a quirkish way, this had acted to preserve Alaskan native land claims long after American Indian tribes, legally or not, had been led to cede over to the U.S. vast areas of land they had long occupied. The treaty of cession from Russia to the United States did not give new citizenship to Alaska's natives. But it didn't vacate their land rights either. The Organic Act of 1884, though, directly defined those claims: "Indians or other persons in said district shall not be disturbed in the possession of any lands actually in their use or occupation *or now claimed by them*, but the terms under which such persons may acquire title to such lands is reserved for future legislation by Congress."

Alaskan natives had been demanding this legislation since

statehood. The state was getting ready to select its own 103 million acres. The natives wanted their land first. By 1968 they had filed claims to 300 million acres—nearly the entire landmass of Alaska, and including the North Slope, which the state also claimed. Interior's Udall now put the freeze on state land patents, and Interior drafted the first native claims bill in 1967. It was rejected by the newly organized Alaska Federation of Natives. The AFN and then-Governor Hickel set up a task force to seek a settlement. The formula that came out of this was the first specific offer to the natives of 40 million acres of Alaskan land and a participation in the state's future mineral royalties.

In 1969 the Federal Field Committee for Development Planning in Alaska added the concept of a $1-billion cash settlement along with the land. This was refined by the AFN to a $500-million cash settlement and a 2 percent overriding mineral royalty, in perpetuity, on state and federal lands. The AFN had another new idea: to set up twelve native corporations, geographically around the state, to supervise each region's land and cash awards.

Congress from 1969 on had wrestled with claims bills that combined these features in dozens of ways. Complex forces played over the deliberations. Alaska's government and its congressional delegation wanted a settlement to unblock the pipeline and get the oil and its royalties flowing. The oil companies pitched in to help obtain a settlement. They wanted to get started too, of course. But they did not favor proposals tying the pipeline corridor in with native claims, for then Congress would have a third jurisdiction over the project, added to the ones that Interior and the courts already had.

Alaskans themselves, typically, were of mixed minds. Even the natives were divided. The moderates in their leadership saw the emerging settlement terms as a longtime dusty dream at last coming true. But some of the more militant leaders, especially in the Arctic Slope Native Association—a splinter group formed where the oil was—felt that if this much of a settlement was being readily offered, the natives had the leverage to hold out for more. Some of Alaska's nonnative people, on the other hand, said that the natives were already being given too much. And over it all, in these years, the looming national sense of coming energy problems was putting a higher and higher priority on Prudhoe Bay's shut-in oil reserves.

Finally, on December 13, 1971, a Senate-House conference

committee compromised on a bill that was acceptable, if begrudgingly in some cases, to everyone. The next day, it passed both houses of Congress by wide margins. The Alaska Native Claims Settlement Act of 1971 gave the natives the right to select 44 million acres of land. For relinquishing any further land claims, they were to receive $462 million over an eleven-year period, plus the 2 percent mineral royalty until they had been paid an added $500 million. Each individual village could be incorporated within the twelve regional native corporations the act created, and a thirteenth corporation was to be set up for Alaskan natives living away from the state. The land freeze would now be lifted. ANCSA was signed into law by President Nixon on December 18, 1971.

7.
1972...The Legal Logjam

Passage of the native claims settlement got the Alaska pipeline over its first and perhaps biggest obstacle. But the environmental lawsuits and federal injunctions against the project still stood. On March 20, 1972, the Interior Department issued the final version of its environmental impact statement. The original 246 pages had now become nine thick volumes, and dealt not just with the environment but analyzed the economics of the project as well. Interior's formal decision on the construction permit was to come after a forty-five-day waiting period to allow comment on the impact statement. During this time, the Environmental Defense Fund drew up its own impact statement countering the government's analysis, and this was delivered to Interior on May 4.

Environmentalists and others opposing an Alaska pipeline were now centering their arguments on a different issue. Interior, they charged, had not adequately considered the "Canadian alternative." What was the Canadian alternative? It was an almost completely different pipeline route that had been proposed almost as soon as North Slope oil was found. It would carry Prudhoe Bay oil eastward across the North Slope, passing through or around the

Arctic National Wildlife Range in Alaska's northeast corner. It would then cross the Canadian border into the Mackenzie River Delta in the upper north of Canada's Yukon and Northwest territories, gathering in any oil in production there. Jersey Standard's Imperial Oil affiliate in Canada had already found oil in the Mackenzie area in 1970, and other Canadian and United States companies were actively exploring.

The Canadian route would then run southeast into the province of Alberta, where it would meet Canada's existing pipeline systems. Here the oil might be sent in two directions—some of it westward to Puget Sound, where it could be transshipped to the rest of the U.S. west coast, and some of it eastward to pipelines and markets in the midwest. This last part of the scheme was favored not only by environmentalists but by a number of midwestern senators and congressmen, who argued that Alaskan oil was needed as urgently or more so in their states as it was on the west coast where the tankers would deliver it from Valdez.

The main environmental arguments went like this. A combined Alaska-Canada route would avoid disrupting all but the most northern part of Alaska. It would establish a route that could be used later for a second pipeline to bring natural gas from Prudhoe Bay and the Mackenzie Delta south to the same U.S. and Canadian markets. The route would cross more hospitable country in Canada than in the length of Alaska. It would cross no earthquake faults. It would avoid the need for a tanker port on the sensitive fishing waters of the Gulf of Alaska, and for constant tanker voyages up and down the Alaskan and Canadian coasts.

There was disagreement about all of this. Arctic and sub-Arctic terrain were the same, many said, whatever side of the border they lay on. The Alaska-Canada route, some 2,600 miles long, would disrupt more than three times as much land. The tanker argument was weakened because tankers would still be used from Puget Sound down the rest of the U.S. west coast. In any case, foreign tankers bringing oil to the U.S. from abroad would still be using routes and ports along the U.S. coast. Most importantly, said environmentalists who were most familiar with Alaska, almost any route that crossed the North Slope into Canada would have to violate the Arctic National Wildlife Range in some way. Even going north along the coast of the Beaufort Sea meant exposing a pipeline to the unpredictable polar ice pack. And any pipeline in that area, they felt, was a first step toward a more ominous risk: a move to open up the wildlife range to oil and gas exploration.

There were other arguments against the Canadian route. The much longer pipeline would take more time and money to build, which meant yet another delay in adding Alaskan oil to U.S. domestic supplies. And Canada, too, had unsettled native land claims. Canadian native groups had watched with interest what their Alaskan brothers had achieved with ANCSA the year before. They could almost inevitably be expected to launch a similar challenge.

There were also some unpleasant questions about conflicting national interests in the U.S.–Canadian pipeline. Did the U.S. want its Alaskan oil dependent on a foreign route for transport? Relations between the two countries were reasonably benign. But Canada had objected after the *Manhattan* voyages to the possible violation of its northern territorial waters. Later in the 1970s, there would be Canadian restrictions on U.S. business and investments there, and increasing limits on oil exports to the U.S. as Canada's own oil supplies dwindled. There would also, though, be a pipeline treaty between the U.S. and Canada. And, in fact, the energy economies north and south of the 49th parallel had become so intertwined that neither country was apt to try to strong-arm the other.

In any case, for whatever combination of reasons it was rejected, the Canadian route for North Slope oil had not been ignored. As it has developed more recently, too, some Alaskan oil may yet find its way from the west coast through Canada's southern pipelines to markets in the American midwest. And Alaskan gas may yet use at least a partly Canadian route to reach the same western and midwestern markets.

On May 11, a week after the forty-five-day waiting period had ended, Interior Secretary Morton announced that he had decided to approve right-of-way permits for the Alaska pipeline. His statement compared at some length the Alaskan and Canadian routes. A line through Canada, Morton said, would cross more permafrost and cause more damage to the terrain and the environment. A line through Alaska ran greater risks from earthquakes and tanker traffic. Morton felt these risks were being protected against. He also said the Canadian line would take "at least three to five years" longer to complete, and that its planning and financing were not as well along as the U.S. project. And, said Morton, it was in the "best national interest" of the U.S. to have "a secure pipeline located under the total jurisdiction and for the exclusive use of the United States."

It looked now, in mid-1972, as though the pipeline's legal progress was accelerating. Only the injunctions were still in the way. The environmentalists were pressing for a judgment that would make the preliminary injunctions permanent. On August 15 Judge Hart called everyone into court. His ruling in the suit by the Wilderness Society, the Friends of the Earth, and the Environmental Defense Fund was in favor of the defendants: the Department of the Interior, Alyeska, and the state of Alaska. Hart's opinion was that Interior had now met all legal requirements for the permits. He dissolved the preliminary injunctions against construction.

The environmentalists, now joined by Canadian groups, immediately appealed the ruling to the U.S. Court of Appeals in Washington. The Appeals Court met to hear the case on October 6. As the judges saw it, at least two points were at issue. Judge Hart had ruled that the environmental requirements of NEPA had been satisfied. But that still left hanging Alyeska's request for a right-of-way wider than the fifty-foot limit imposed by the Mineral Leasing Act of 1920. Alyeska was seeking a special land use permit to enlarge the right-of-way, a type of permit that pipeliners had had no trouble getting in the past. The appeals court judges adjourned to consider this. It would be early the next year before they reached their decision.

In Houston, Alyeska's design was also making progress. The design review group of pipeline engineers and U.S. Geological Survey scientists had continued to analyze the field study data as they came in, and the technical debate over the pipeline's engineering had resolved itself into a new guideline for the project. It was to be designed from the start to be as fail-safe as possible and, with Interior's sixty pages of stipulations as the construction rule book, it was to be built that way. There would be, the government scientists said, no going back in mid-construction to correct problems that were cropping up, as pipeliners were used to doing. In Alaska the problems were to be foreseen and headed off before anything was built in the first place.

"In other words," says Vernon Cardin, the manager of engineering then, "you were to lay the pipeline where repairs would

not be required, and that meant you designed it so that nothing
. . . the environment, or the pipeline . . . would ever have to be
repaired from Day One. It was a revolutionary concept of design
. . . that you did not consider repair."

What it also meant was more elevated pipeline. "With no
repairs permissible, the project automatically swung to more pipe
being above ground," Cardin says. "Pipe could be buried if . . .
and only if . . . it were in terrain that would be just as stable after
thawing as it was in a frozen condition. Naturally the alternative to
that was to lay the pipeline on top of the ground."

By 1972 TAPS and then Alyeska had already spent three years
classifying several ways, or "modes," to build portions of the line in
different terrain conditions: thawed soil; bedrock; thaw-stable sand
and gravel; and the less stable areas where special burial methods
or elevated construction might be needed. Working with the
USGS, Alyeska and its engineering consultants had now arrived at
five main construction methods: conventional burial; burial on
gravel pads; elevation on pilings; special burial, possibly refrigerated; and either burial or elevation in river floodplains.

The above-ground design progression was from simple ideas
to increasingly complex ones. "One was just laying the pipeline on
top of the ground," Cardin relates. "Another one was putting it on
a series of skids on the ground. And another was to build a berm
that would get the pipeline up considerably above the ground, but
still provide continuous support for the pipe. At that point, we felt
that putting a pipeline on top of pilings at intermediate spaces was
not ideal from an engineering standpoint."

It hadn't been done before, not in cross-country crude oil
pipelining. In an oil refinery or petrochemical plant, piping to
move feedstocks and products over short distances around the
plant is elevated for easy inspection and maintenance. But the only
elevated long-distance oil line to date had been Tapline in the
Middle East, raised a foot or so over the desert on firmly positioned
concrete beams.

This wasn't possible in Alaska. Nor was any other method that
allowed contact between the pipeline and the unstable soil it was
being raised above. "What outlawed all that was the heat of the oil,
which would be transported down into the ground, or through a
gravel berm into the ground, and the permafrost would start melting," Cardin says. "You had to have air circulation under the
pipeline to carry away the heat and keep the soil from disappearing

on us. So it was really the heat transfer from the pipeline that caused us to go to pile supports exclusively in the above-ground design."

The next design decision was what kind of supports to use. "Our original concept was wooden piling, because it transferred less heat than concrete or metal piling," Cardin recalls. "But timber pilings were not strong enough to supported the concentrated loads and stresses of the pipeline. Concrete was not flexible enough. So it worked out in the end that we had to have metal pilings after all." It also put Alyeska back into finding a solution to the original problem: heat transfer.

Moreover, in 1972, the problem had grown. Alyeska's first project description in mid-1971 had identified 136.1 miles of line to be elevated. As the description was later refined with new data from the field, it appeared that a total of 423.7 miles of the route lay in soil conditions that would require some kind of above-ground construction or special burial methods. A large part of the new above-ground mileage was in the Copper River Basin, where soil instability had finally ruled against burying so much of the line. At the start of 1972, then, about 312 miles of the pipeline were still marked for conventional burial. Another 150 miles or so were to go up on gravel berms. Almost 200 miles were now designed to be put on pilings. That left some 70 miles of refrigerated burial, and a few miles of special above-ground construction in the more difficult river and floodplain areas.

But the berms, as Cardin said, were still likely to cause heat transfer problems. And refrigerated burial was costly, hard to engineer, and hard to build. So the effort now in 1972 was to improve and if possible simplify the several different modes of above-ground construction. The research in Houston on a thermal piling system was far enough along that Alyeska and its consultants felt confident of substituting this technology for most of the special burial sections. More of the special burial was eliminated by development of a deep burial method that would put the pipeline well below the level of unstable permafrost. And plans were made now to eliminate almost all of the gravel berms by raising those portions of the line completely off the ground, on pilings. It would be decided later, section by section, if the pilings would be thermal or nonthermal.

As 1972 wound down, Vernon Cardin was leaving the project to return to Humble Pipe Line Company. The company, though, was about to change names. Standard Oil Company of New Jersey decided in November to bring all of its worldwide operations under one new corporate label. Jersey Standard itself became Exxon Corporation. Humble Oil & Refining Company, its historic Houston-based domestic arm, became Exxon Company USA, Humble Pipe Line became Exxon Pipeline Company.

The pipeline design, as Cardin left the engineering staff, was moving beyond the concept stage. "We were well along on all of the facilities by then," Cardin recalls. "We had all our concepts established, and our criteria and procedures . . . and we were in the process of reviewing all this with the government. It seemed like some of the applications of those concepts were changing daily. But the basic design was established."

The next task would be the detail design. "We had a part of that design already done and supported with all of the documentation that was required," Cardin says. "I think we were calling it a 'mile-by-mile' design . . . but that was a misnomer. The refinements came along all the way through, right up until today. We had to design that thing foot by foot."

8.
1973...Authorizing the Pipeline

By early 1973, the North Slope oil discovery was five years old. Some of the 800 miles of pipe had been on the ground in Alaska for more than three years. On February 9 the U.S. Court of Appeals issued its decision. The court partially overturned Judge Hart's ruling to let the pipeline go ahead. No judgment was made on whether the project now conformed to the National Environmental Policy Act. But the court said it had no jurisdiction over the right-of-way limits in the Mineral Leasing Act of 1920. Congress had created the act, and although Secretaries of Interior had been granting exemptions for years, Congress would have to change it to allow a right-of-way of more than fifty feet. The Justice Department appealed this decision to the Supreme Court, which declined to review it.

So the pipeline was in the hands of Congress after all. Both the White House and the Department of Interior urged quick congressional action on a law to clear the pipeline permits. Within weeks, a flurry of bills sailed into the hoppers of the Senate and the House of Representatives. True, the Interior committees of the two houses had already held hearings on the project. But that was in 1969. The

87

design had barely begun. And this was before the passage of NEPA, before the native claims issue arrived in Congress, before the proposal of an alternate route through Canada. And, before the "energy crisis."

Was the energy crisis invented to hurry up the Alaska pipeline? Clearly, the Nixon administration was pushing to get the project started. It's been charged that the oil companies on the North Slope deliberately underestimated Prudhoe Bay's reserves to reinforce the idea that U.S. oil was running low. But it's also been charged that the oil companies, to justify the pipeline, pointed to their huge Alaskan finds as a major solution to the growing U.S. oil shortage. It can't have been both ways. Atlantic Richfield was freely saying as early as mid-1970 that the Slope might hold as much as 12 billion to 15 billion barrels of reserves. Standard Oil of New Jersey had said at the same time the reserves might be as high as 20 billion barrels.

In 1973, as the pipeline legislation began to take shape, energy shortages were becoming very much a theme of the times. Warnings from the oil companies themselves had begun long before that. However bullishly or bearishly the oil industry was estimating its reserves, the fact was that the rate of finding new oil in the U.S. was declining. Exploration and development costs were rising steeply. But prices in the marketplace were not. Industrial oil and gas users, and consumers of fuel at home and on the highways, were paying no heed at all to rising pleas for conservation.

Apart from the oil industry, the debate over adequacy of energy supplies was spreading among other experts. The USGS and other government agencies, and the National Academy of Sciences in the private sector, were scaling down their estimates of U.S. oil and gas reserves. There was also deep concern about OPEC's increasing aggressiveness—and effectiveness—in setting higher world oil prices. And, in 1973, within the State Department and its foreign service corps, there were mounting fears of a new war in the Middle East. No one was clairvoyant enough to predict that a war would come before the year was out. But the danger of it was building up alarmingly.

All of this helped to set the climate for a pipeline bill. The step-by-step deliberations were keyed a good deal more closely to the familiar issues of the project. NEPA, the right-of-way limits in the mineral act, and some foreign policy questions closer to home were the main focal points.

Anyone who has traveled along the route of a modern pipeline under construction will know that the fifty-foot right-of-way limit in the Mineral Leasing Act of 1920 was badly out of date. Nor was it actually adhered to in recent projects. Pipelines early in the century were much smaller, and the "equipment" used to build them up to 1920 was likely to be mules and wagons in contrast to the thundering heavy machinery needed for today's large-diameter lines. The Alaska pipeline's request for one hundred feet of right-of-way, even if it was twice the statutory limit, was simply a long overdue catching up with construction realities. Superhighways and modern bridges or air terminals can't be built by 1920's standards either.

Section 28 of the Mineral Act gave the Secretary of the Interior jurisdiction over pipeline rights-of-way. Congress decided to let the fifty-foot limit stand. but it amended that section to allow the Secretary to increase a pipeline right-of-way temporarily during construction, and also to allow a wider right-of-way if the pipeline's operation and maintenance required it. The decision had to apply to a specific project.

In this way Congress met the legal question over the right-of-way. The environmental questions were a tougher problem. The Court of Appeals had not ruled on whether NEPA's requirements had actually been satisfied. What worried Congress now, as well as Interior and the pipeline company, was that just passing legislation to authorize the pipeline would not be enough to clear away the lawsuits still pending against the project. These had now been in the courts for more than three years. None had yet been settled. Modern judicial review and its careful appeal process were bona fide constitutional safeguards. At the same time, though, plaintiffs in a litigation could use the courts and a law like NEPA to keep a project like the pipeline tied up in knots for years.

Congress was in a dilemma. It had enthusiastically created NEPA in 1969 as a response to nationwide environmental sentiment. The Alaska pipeline was the new law's first major test. NEPA was working effectively to do what it was meant to do. In this case and many others since 1969, it was revolutionizing the careful documentation and process of approving big projects. As to the pipeline itself, everyone but the most fiery-eyed of the environmentalists agreed by now that the Alaska project conformed to NEPA's intent. What, then, to do about the problem of the endless legal delays?

Alaska's own senator, Mike Gravel, came up with an answer that almost no one would have thought was even a practical possibility. Gravel proposed an amendment to the pipeline bill that would limit and compress any further judicial review of the project. Permits granted under the new law for the pipeline and the haul road were not to be subject to any more challenges under NEPA. The Gravel amendment was almost as controversial as the pipeline itself. Critics charged that it was cutting the heart out of NEPA. Gravel and its other sponsors said that it was simply putting a time limit on the challenges NEPA had already made possible. The amendment came to the Senate floor on July 17. It exposed Alaska to another of those excruciating close votes that had decorated Alaskan history for more than a century. The vote was a 49–49 tie. By tradition the Senate's presiding officer, Vice-President Spiro Agnew, broke the tie—in favor of the amendment.

Editorials in major newspapers continued to score the Senate's action for "bypassing NEPA" and "exempting the project from NEPA." *Business Week* editorialized that this was "exempting the pipeline from further challenges under the 1969 law." Said the magazine: "The amendment declares the pipeline environmentally acceptable and not subject to further judicial review. . . . Exempting the pipeline from NEPA would set a dangerous precedent." Now, the critics said, waivers would be sought on all projects protected by NEPA. And the pipeline law itself was sure to be challenged because Congress had no constitutional right to interfere in judicial processes.

None of this was true. The amendment did not allow the Alaska pipeline to "bypass" NEPA. It has not allowed any other project since then to avoid review under NEPA. The point of the amendment was that the pipeline had *already* satisfied NEPA's provisions. Interior had already prepared the exhaustive impact statement that the law said it must. The force of that statement on the project has stood firm. And impact statements have kept on playing a key role in project approvals ever since.

Nor was the Gravel amendment meant to bring an abrupt end to litigation, and it did not do so. The law, in fact, set up a sixty-day period after the right-of-way permits were granted, when additional lawsuits could be filed. The important difference here was that a time limit was set. Any new cases were to be handled quickly. A special panel of three federal judges was created to hear them. And any appeals of their decisions were to go directly and immediately to the Supreme Court.

As it turned out, the disputed amendment did not lead to a new wave of legal action. The pending lawsuits were vacated in an orderly way in the months after the pipeline authorization act passed. No new cases were filed. With the amendments to the Mineral Act and the congressional endorsement of Interior's environmental and technical review of the project, the pipeline's opponents were no longer objecting that due process under the law had not been met. "Any further judicial review would have been only with respect to constitutional questions," says John D. Knodell, the Exxon USA attorney who was Alyeska's general counsel and manager of law and government relations. "There were no challenges . . . and I have always assumed that this was because no one could find any denial of constitutional rights."

In the congressional debate, before the bill came to a vote, there remained only a couple of "foreign policy" questions. Amendments by Senator Walter F. Mondale of Minnesota and others to hold up the Alaska project while the Canadian alternative was studied again were defeated. There was also a concern that part of Alaska's oil would be sent into export markets—in particular, Japan—because the U.S. west coast would not be able to use it all by the time the pipeline was built. A provision was put into the bill that this could be done only if the President approved it in the national interest at the time. Congress, moreover, could reverse the President's action.

What was the legacy of the years that went into gaining permission to build the Alaska pipeline? The long and complicated approval process was extraordinary treatment for an industrial project. Did the pipeline, as a lawyer would put it, "make new law?" Or was it, as the pipeline's engineers say of their own work, an "extending" of established legal technology—pushing existing law into new ground?

The pipeline clearly was the catalyst for some legislation that came very close to "new law"—a long jump, at least, beyond past measures of the same sort. Some of this was historical accident. The pipeline at the start was an ambitious but otherwise straightforward business undertaking. But it converged with some powerful social and economic currents of its times.

The best example was the native claims settlement. The native minority movement, best known for the protests of the American Indians, was making only the slow progress toward recognition of its rights that had characterized every minority movement. Alaskan natives had clung to their claims for centuries. They might have waited decades longer to achieve what the pipeline did for them in less than two years. In a single law—ANCSA—a group of American natives gained 44 million acres of what native minorities, nation-wide, hope ultimately will be a 100-million-acre settlement of their land rights. They gained almost a billion dollars besides. And they carved a path for native ambitions everywhere.

The pipeline also converged with the landmark environmental law of its day. No one planned it that way. The project was not the catalyst for NEPA. But it was surely its proving ground. Reinforcing this was the locale where the pipeliners wanted to build, for the aura of Alaska was enough to inspire strong emotion all of itself. NEPA, in Alaska, has rewritten the rule book for big projects on federal lands. For better or worse—depending if you're talking to a pipeliner or an ecologist—it gives government an unprecedented amount of control and surveillance over projects from now on. And the Alaska pipeline has been a casebook of experience in using that power.

Quinn O'Connell of the Washington law firm of Connole & O'Connell helped the pipeline in its efforts to get permits almost from the start. "Unique" is the blanket word he uses to describe each step of the way. The issuing of detailed environmental and technical stipulations was unique in itself, O'Connell says. Issuing them for one particular project was also a new event. Then there was the twenty-seven-page right-of-way permit, with another thirty pages or so of supporting material, including a summary of the stipulations. The pipeline was also the first private project to be granted a materials priority under the Defense Production Act of 1950. That law had been created for government projects, but it was applied to the pipeline when it was feared that certain grades of steel and other needed items might be in short supply.

"And, of course, the Trans-Alaska Pipeline Authorization Act itself is unique," says O'Connell. "It is also, I think, one of the best-written pieces of legislation to come out of the Congress. Because it is so well written, it has been possible to build the pipeline." It was unusual enough that a law had to be passed at all. But O'Connell then points to some of the specific features.

The pipeline act requires the Interior Secretary to see that the project sets up and conforms to an Affirmative Action program for minority hiring. Such programs have been a part of government-sponsored projects for some time. The pipeline was the first private project required—by name and by a specific law—to have this kind of program, and to enforce it through the contractors that Alyeska in turn hired to do the work. In its permit, too, Alyeska was pledged to submit a plan to the Interior Secretary whereby it would recruit, train, and hire Alaskan natives to fill jobs on the project.

Another feature of the act is that it not only gives the government a strong surveillance role over the project but also requires the pipeline company to reimburse the government for the costs of that surveillance. Those costs had added up to $12.3 million by the time the permits were granted, and Alyeska has paid about another $47 million during construction. Then the act sets up an oil-spill liability fund, to be maintained by Alyeska during construction and as long as the pipeline continues to transport oil. "The liability applies," says Alyeska's John Knodell, "on land or on sea . . . so the fund still applies to Alaskan oil even if it is spilled in California waters."

Thus the pipeline law and its permits reach out beyond Alaska. Technically, the Secretary of the Interior's jurisdiction in the past has not gone past the territorial limits of U.S. coastal waters. The Alaska pipeline, in effect, extends his authority out to sea. The permit prohibits vessels coming to Valdez from discharging their oil-contaminated ballast before they arrive. Alyeska itself was already refusing to allow any ship—U.S. or foreign—to berth and take on oil at Valdez unless it brings its ballast with it to be treated in the terminal's facilities for this purpose. "Normally, the Secretary of the Interior would not have any right to challenge these vessels," says O'Connell. "This is using a right-of-way permit over federal lands to accomplish a result out in the middle of the ocean."

It was Valdez, in fact, that introduced Quinn O'Connell to the project. Atlantic Richfield had called him about a problem with the first permits. "They needed to get the Department of Interior to give approval to the state of Alaska for the terminal site, and they couldn't figure out why it was taking so long," O'Connell says. "Five years later, I said: 'Remember that little problem I went down to ask about in 1969?' "

Five years later; Alaskan oil and its pipeline now had a new urgency. On October 6—Yom Kippur—Egypt and Syria invaded Israel. On October 18, in retaliation for United States military aid to Israel, Saudi Arabia and other Arab members of OPEC declared a total embargo on oil shipments to the U.S. Arab oil production was also to be cut back, and by early 1974, the world price of oil would be quadrupled by OPEC to more than $11 per barrel.

The reserves at the North Slope were needed now. Congress wound up its debates on the pipeline bill within a couple of weeks, and the usual conference committee of the two houses quickly worked out the differences between the House and Senate versions of the legislation. The Trans-Alaska Pipeline Authorization Act of 1973 was passed by the House of Representatives on November 12 with 361 votes in favor and 14 against. The Senate passed it the next day by a vote 80 to 5. President Nixon signed the bill into law on November 16.

9.
1974...Under Way

In the half-light of the early Alaskan winter, the Yukon River was a wide band of frozen silence. On the south bank the crew from Fairbanks was climbing down from the trucks. It was December 1973. The temperature today would hover between 20° and 30° below zero. But that's normal winter working weather in north-central Alaska. You just dress well for it, and you never turn off the motors on your equipment or it will freeze up.

The work today was to start building another ice bridge across the Yukon. Several weeks later, when the bridge was done, heavy trucks, bulldozers, and other construction machinery could move north of the great river into the mostly trackless southern foothills of the Brooks Range. Here the crews would smooth out a temporary 360-mile snow and ice road along the route laid out by survey parties in past years. Up this winter road would follow battalions of men, machinery, and supplies bound for the shut-in camps located every fifty miles or so all the way to Prudhoe Bay. In the spring this small army of several hundred workers would start laying down Alaska's first permanent secondary highway north of the Yukon.

The highway would one day belong to the state. But first it would be a construction haul road.

The building of the Alaska pipeline had begun.

Four years had gone by since the first mobilization over the Yukon, abruptly cut short by the long lawsuits. The pipeline's cost, Alyeska had announced the previous summer, was up to $4.5 billion. The long-awaited federal right-of-way permit was issued by Interior Secretary Roger Morton on January 23, 1974. Now there was a scramble to organize as much of the construction effort as possible before the spring breakup. It was pure accident that the authorization came as winter was over Alaska. It's another of those Alaskan paradoxes that moving equipment around like this is best possible when everything is frozen solid. Construction itself goes faster when the state thaws out. As earlier projects in the High North had proven, though—and as this project would prove again—an amazing amount of work can also be done in the dead of an Arctic winter.

The ice bridge and the snow road did not have to wait for the permit. They were nearly finished when it came. The logistics of the northward flow that began now would dwarf heroic episodes of the past like the Berlin airlift and some great overland efforts in Alaska itself: the Gold Rush in the 1890s, the military movements during World War II.

In eighty-three days, from late January to mid-April, a force that at one point reached 680 workers moved some 34,000 tons of machinery and materials into northern Alaska. This took 671 aircraft flights and 1,285 trips by truck. Seven mothballed construction camps were opened and enlarged, and five new camps were built. Five temporary airstrips were built over the snow and ice, to be replaced in spring by a permanent gravel-based runway at each camp. The crews kept at it around the clock in temperatures that dropped as low as minus 68° at Prospect Creek camp, about twenty miles above the Arctic Circle, the site a few years earlier of the coldest temperature ever recorded in the U.S.: minus 80°. Nor do you reckon without the wind-chill factor in Alaska winters. This year, the work went on in wind-chill temperatures of 115° below zero.

What does it take just to get ready to build an Alaska pipeline's supply road? Start with 900,000 gallons of diesel fuel—400,000 delivered by road, 500,000 by air. To use the fuel, bring up 716 construction vehicles and other pieces of equipment from below the Yukon. Take another 75 pieces of equipment out of mothballs at Prudhoe Bay, where it's been stored since 1970, and completely recondition it. Bring up 600 tons of replacement parts. Bring up 600 prefabricated camp buildings, and 2,200 tons of camp supplies.

Do all of this in less than three months. Okay, now you're ready to start on the real work.

For the planning and civil engineering of the haul road, Alyeska named Michael Baker, Jr., Inc., a firm based in Beaver, Pennsylvania, as its subcontractor. The highway bridge across the Yukon River was being designed by the state of Alaska itself, and Alyeska was sharing in its $32.2 million cost. The contractor picked to build it was Manson-Osberg-Ghemm, a joint venture of Manson-Osberg Company and Ghemm, Inc. The steel parts for the bridge were being made in Japan, where they would be test-assembled by the fabricators and then taken apart again for shipment by sea to Alaska.

The haul road route was divided into eight sections. Each two portions would be built north and south from a center point until they all connected. Bids were invited from nine construction firms; and on April 5 four were picked as execution contractors for the road. Green-Associated and General-Alaska-Stewart were both joint venture companies. Burgess Construction Company was the firm that had built the 53-mile Livengood-to-the-Yukon road in 1969. Morrison-Knudsen Company, Inc., from Boise, Idaho, was a worldwide heavy construction specialist.

Each stage of the project now was a matter of more men, more materials, more time—more of everything than the stage before it. Haul road construction started on April 29. At the peak of the effort, Alyeska and its contractors had more than 3,400 workers deployed over the route. The supply flights during the winter had been just the beginning. Now a squadron of more than sixty aircraft, ranging from helicopters to big, fixed-wing transports and air tankers, was crisscrossing the skies over northern Alaska in support of the road-building. More than 127,000 flights were made, an average of about 700 a day. Eight and a half million gallons of fuel were flown in to power the construction equipment and the camps. Another 160,000 tons of supplies and materials were also transported by air. And, by early summer, barges were being used to take materials directly by sea to Prudhoe Bay.

A secondary road in the north of Alaska is basically a sophisticated gravel road, highly compacted and graded to strict specifications. Only a gravel overlay like this would protect the underlying tundra in the far north, the less vulnerable taiga and muskeg ground cover in the sub-Arctic, and the subsurface permafrost that might be found anywhere along the route. And only a well-graded gravel road would stay there, standing up under the constant and punishing traffic of heavy trucks and equipment. Alaska's only road link to the Lower 48—the Al-Can Highway that was pushed through British Columbia and the Yukon Territory in World War II—is a difficult, no-nonsense gravel road over most of its long Canadian mileage. Only after it crosses the Alaska border is it now paved as part of the state's main highway system to Anchorage and Fairbanks.

Gravel is a standard building material in Alaska because there is so much of it. It is found in abundance in the state's endless hillsides and stream beds. Environmentalists may argue the point, but Alaskan gravel is a finite resource only in the sense that every natural resource at last is finite. Estimating how much gravel there is in Alaska is about as sensible as calculating how much sand there is in the Sahara. For the pipeline haul road, trucks hauled more than 31 million cubic yards of gravel and a million cubic yards of rock to one point or another of the route. Gravel was also to be used as a terrain-protecting foundation for most of the buildings along the pipeline, and as a bed for much of the buried steel pipe.

The environmental stipulations went into effect with the first truckload of gravel. Alyeska and the contractors were pledged to leave hillside gravel pits in good enough shape that erosion would not follow the next turn of the seasons. Gravel had to be mined carefully in dry stream beds so as not to alter the watercourses or endanger fish and other wildlife using the streams. Alaska's wild creatures got priority again and again. Early in the road work, a crew was sent elsewhere until a nearby hibernating bear woke up. At Franklin Bluffs on the North Slope, the route of the road was shifted and the site of the construction camp there was moved because peregrine falcons were nesting in the area. That summer, road work near Chandalar, high in the Brooks Range, was put off because a colony of Alaska's Dall sheep was lambing there.

Beyond these singular events, pesky delays and ad-libbed shuffling of plans were to be expected on a large, fast, and complex construction effort. Equipment would break down, or not show up where and when it was supposed to. Supply shortages would develop at the worst possible time. Gravel beds were not always found where the most gravel was needed. And more gravel was used overall than had been estimated—13 percent more, or an added four million cubic yards. It was found, somewhere, and moved to where it was needed. Alaska's summer weather was known to get almost as extreme as its winter climate, and contractor crews in the Brooks Range foothills sloping north to Prudhoe Bay this summer worked in temperatures of 90°—*above* zero. The crews at the different camps strung out along the road competed boisterously with each other. Each crew kept daily scores of its progress and how this stacked up against the work being done from the other camps.

Birds, beasts, and equipment or gravel problems notwithstanding, the haul road was a roaring start to the project. The initial gravel overlay for the entire 360 miles was finished on September 29. It had taken just 154 days. More than three million man-hours had gone into it. The final linkup, complete with a ribbon-cutting ceremony, took place on a rangy plateau about 100 miles north of the Yukon, between the camps at Prospect Creek and Coldfoot. Twenty permanent bridges also had to be built this year across streams on the route. More than half were completed by fall, and work on others kept going while temporary bridges were installed so that the new road could be used from one end to the other. At the site of the Yukon bridge, one of the five big concrete piers was already finished and two others were well along, while work on the north bank abutment had begun. Final grading of the road continued until it was done on November 15. The haul road was, of course, the first American highway to cross the Arctic Circle.

That spring, as the road work got under way, Alyeska had picked the contractors who would directly supervise the sprawling pipeline project itself.

The construction plan of the pipeline system, like its design, was divided into three main parts. There was the mainline—the 800-mile steel tube from Prudhoe Bay to Valdez. There were the pump stations—twelve of them spotted along the route, with nine to be finished for the first phases of operations while site preparation was done at the other three. Then there was the marine terminal at Valdez. The huge onshore oil storage, processing, and tanker port had to be virtually done by the initial start-up date of mid-1977. That meant completing eighteen of the eventual thirty-two storage tanks in the two big tank farms, and four of the eventual five tanker berths, with all of the docks and offshore loading machinery for the seagoing tankers. It also meant finishing the tanker ballast treatment plant, the oil vapor recovery plant, the terminal's power plant and other service facilities, and the entire pipeline system's computerized control center.

On an ordinary project, a company like Alyeska would probably have chosen a single prime contractor who would then subcontract smaller parts of the work to other companies. The scope of this project was too vast to do it that way. Alyeska instead named two construction management contractors—CMC's—each with long experience in building massive plants and other projects around the world. Bechtel Corporation, based in San Francisco, became the CMC for the mainline. Fluor Alaska, Inc., a unit of Fluor Corporation of Los Angeles, which had been working on the terminal engineering since 1969, was made the CMC for both the terminal and the pump stations. Fluor Engineers & Contractors, Inc., would also do much of the construction work on the terminal and stations.

The next step was to pick the execution contractors to work under Bechtel to lay the mainline. From south to north, the route again was cut up, this time into six sections. The length and boundaries of each was decided by the range and difficulty of terrain it passed through. The list of contractors, mostly teamed up in joint ventures, could be read like a roster of the international construction industry.

Section One ran 153 miles from Valdez north to Sourdough. The execution contractor was the River Construction Corporation, a division of Morrison-Knudsen, and quickly to be known on the project as "MK-River."

Section Two covered the next 149 miles from Sourdough up past Delta Junction. The contractor was Perini Arctic Associates, a joint venture of Perini Corporation, Majestic Construction, Wiley Oilfield Hauling Ltd., and McKinney Drilling Company.

Section Three was 144 miles, from Delta past Fairbanks almost to the Yukon. H. C. Price was the contractor, another four-way joint venture between H. C. Price Company, R. B. Potashnick, Codell Construction Company, and Oman Construction Company. The venture soon went under the tag of PPCO.

Section Four was 143 miles, from south of the Yukon to Cold-foot. The execution contractor was Associated-Green, the same venture of Associated Pipeline Contractors, Inc., and Green Construction Company that was already at work on the haul road—with the names reversed.

Section Five took the route from Coldfoot to Toolik, and Section Six went the rest of the distance to Prudhoe Bay. Together the two sections covered 210 miles, and they were awarded to the same execution contractor. Arctic Constructors was also a multi-member venture: Brown & Root, Inc., a subsidiary of Halliburton Company, Ingram Corporation, Peter Kiewit Sons, Inc., Williams Brothers Alaska, Inc., and H. B. Zachry Company.

The contracts for all six sections, Alyeska announced, totaled $545 million, exclusive of materials costs. They were fixed fee, fixed supervision, fixed equipment, reimbursible labor contracts.

Since 1970, north of the Yukon, Alyeska already had in place most of the construction camps it would need. Now it built camps along the southern half of the route. Seven were spaced along the way for the mainline workers. Another twelve were placed at the pump station sites both north and south of the river. The largest camp of all, for three thousand workers, was built at Valdez. In total, at this point, Alyeska felt it needed twenty-nine construction camps for the project. They were designed and built almost like small, utilitarian villages, virtually self-contained to provide all of the necessities of life—and, in Alaska's remoteness, some of the genuine comforts as well.

Fairbanks, not quite halfway up the pipeline route, was set up as the project's construction headquarters. Fairbanks has long been Alaska's second-largest city after Anchorage, but in 1974 it was still only a fair-sized town by Lower 48 standards. One of its assets, though, was Fort Wainwright, a large Army base on the outskirts. For the roughly thirteen hundred people Alyeska expected to assign to Fairbanks, as well as for storage facilities, a worker orientation center, and the central offices for the Section Three contractors, the pipeline company contracted with the Army for the use of vacant barracks and office space. It also arranged to use Fort Wainwright's airfield to relieve some of the pipeline traffic pressure that was building up at Fairbanks International Airport.

Contracts were also awarded this summer to companies that were to work under Fluor Alaska at the Valdez terminal. Site preparation was to be done by Morrison-Knudsen. Chicago Bridge &

Iron Company would put up the tank farms, with steel supplied by Nippon Steel, one of the three Japanese makers of the pipe in 1969. The tanker berths were contracted to Kiewest, a joint venture of Peter Kiewit Sons and Willamette-Western Corporation. The piping insulation contract went to General Electric.

Fluor Alaska was also at work on the pump stations. Even while the road to Prudhoe Bay was being built, construction had begun there on Pump Station 1. This was to be the largest and most complicated of the stations, for it was the "origin" station where the oil would come from the field's gathering systems to be metered according to which oil company it belonged before it was sent flowing south. With the ingenuity bred by Arctic engineering, P.S. 1 was being built in a lake bed. Prudhoe Bay is dotted, in its brief summer months, by hundreds of small and large lakes. On the theory that the settled bottom of a lake would be the most stable base for such a critical facility, one lake was drained and then filled in with 460,000 cubic yards of gravel. Mechanical refrigeration in the form of buried piping carrying chilled brine was installed so that the ground would stay constantly frozen all year.

Prudhoe Bay was also the site, this summer where the first four miles of mainline pipe were welded together. This had two purposes. It was used to test field welding and inspection methods, trying out a variety of hand-welding and machine-welding techniques. And, when finished, the four-mile section was used for temporary fuel storage.

Another pipeline test facility was built near Fairbanks. This was a 150-foot section of elevated pipe with one of the hulking 28-foot, 60,000-pound mainline gate valves installed in the middle of it. The section was filled with oil at 140°, and the valve was kept operating through the summer and winter to test its performance over the awesome range of temperatures that Alaska could inflict on a chunk of metal. Fairbanks gets downright hot in summer. Its winters can be quite otherwise. To prove it, the next January, temperatures around Fairbanks dropped as low as 75° below zero—a good workout for a valve or any other piece of pipeline hardware.

One important task had preceded much of the work in 1974. In April, Alyeska had arrived at a broad project labor agreement with the AFL-CIO's Building and Construction Trades Department and with the heads of sixteen international craft unions. The heart of the pact, in view of the project's urgency as it at last got under way, was a provision in Article VII: "During the term of this Project Agreement, there shall be no strikes, picketing, work stoppages, slowdowns, or other disruptive activity for any reason by the Union or by any employee, and there shall be no lockout by the Contractor."

In other words, the pipeline was going to be built from start to finish on a "no-strike" basis. This kind of treaty between labor and management was not entirely new. It had been used in that most massive of government projects, the United States manned space program of the 1960s. On a private project, it had also been used by the contractor that built Disney World in Orlando, Florida. But, says E. G. Sheridan, Alyeska's manager of labor relations, "Up to the time we did it, the notion of a project *owner* making such an agreement was somewhat unique."

Gayle Sheridan was a young Oklahoma lawyer who came to the project from British Petroleum in 1969. With construction expected to start at any moment then, the pipeline's owners were anxious to have work rules firmly set for the thousands of different craft union members that would soon be marshaled into Alaska. It was beyond question that the pipeline would be built by union labor. But the owners felt then that labor relations should be left to the individual contractors, as was customary on such projects. The contractors thought so, too.

During the legal delays, Alyeska had second thoughts. The pipeline's costs were rising yearly, and so was the potential damage of being exposed to strikes and other labor problems once the project did get started. Contractor agreements with the unions were not going to help Alyeska avoid this exposure. It was Alyeska itself that had to have the binding agreement, and it had to foresee any sort of labor problem that might come up. The contractors only reluctantly accepted this. Most of the unions were enthusiastic. Construction was slumping all over the U.S., and the Alaska project offered them a big, 100 percent union job in return for their pledge not to hold it up. They also agreed, although less happily, to the project's own pledge to use Alaskan resident and native labor where possible.

Nothing works perfectly, and there have been labor flare-ups on the pipeline despite the project pact. Still, says Gayle Sheridan: "Our stoppages have never been system-wide, and never for more than a few days, because it's in the agreement that we get expedited arbitration or we get into court immediately." Also, he says, "Our labor rates have been no more than we originally programmed." Most importantly, the degree of labor peace has brought the pipeline to completion on schedule in mid-1977. "In my opinion," says Sheridan, "this is the most effective and enforceable labor agreement ever arrived at, and none has come this close since."

All told, 1974 was a full year. The Alaska pipeline was now in motion. Much had been done all up and down the 800 miles of Alaska that the project would traverse. And much advance work had been laid down for the start of actual pipelaying in 1975.

A big decision had also been made by Alyeska's owner companies in July. The start-up capacity of the line, looking ahead three years, had originally been pegged at 600,000 barrels of oil per day. It was decided now that the system would operate more economically if it aimed at a doubled initial throughput of 1.2 million barrels. The change required some of the companies to rearrange their financing plans, and so there was another realigning of the corporate shares in Alyeska. BP's portion was stated separately again. Now the ownership stood like this: SOHIO, 33.34 percent; ARCO, 21 percent; Exxon, 20 percent; British Petroleum, 15.84 percent; Mobil, 5 percent; Union Oil and Phillips Petroleum, 1.66 percent each; and Amerada Hess, 1.5 percent.

Neither the doubled throughput nor the new ownership alignment had a direct effect on the final cost of the project, although some pump station costs were now encountered sooner. But everything else continued to do so. As the nights grew longer over Alaska in late 1974, so did the bill for the pipeline. The new estimate, announced on October 31, was $6 billion.

10.
The Design Comes
Together...and
So Does Alyeska

The pipeline's price tag was being pushed up each year in just about every cost category. In the five years since the first $900-million estimate, inflation had been raging throughout the U.S. Inflation in geographically isolated Alaska, moreover, always put a premium on everything there. So did Alaskan construction conditions. But beyond all of this, there were the project's technical and environmental costs. The $900-million pipeline in 1969 was based on rather simple and conventional engineering concepts. By 1974 the project had become an engineering spectacular.

Alyeska was now based entirely in Alaska. Its expanding staff was quickly filling up a five-floor office building out on Bragaw Street east of downtown Anchorage, plus a smaller annex next door. During the legislative delays, many new faces had come into the project's top management under Ed Patton. "We had been in a holding pattern, and even the holding pattern had stages," Patton said. "We lost lots of good people . . . if you're not doing anything, it's hard to hold a tiger when there's no excitement."

Now there were lots of new tigers. They came, as the project had been staffed all along, on loan from the major owner companies. Peter DeMay, a New Jersey-born mechanical engineer and chief project manager from Exxon Research & Engineering Company, had helped build refineries and similar projects in Europe and the Caribbean. He arrived at Alyeska in early 1972 as vice president for project management. Charles R. Elder, Jr., came from SOHIO at the same time as executive vice-president for operations. A rangy Kansan, he was a chemical engineer who had worked in pipelining and transportation management, and most recently as marketing vice-president. George M. Nelson, a Missourian who had been SOHIO's manager of management systems, was sent up early in 1974 as vice president for administration.

Some veterans of the legislative years in Washington were also on the Anchorage staff now. Ronald H. Merrett, from British Petroleum, had been a planning manager in the North Sea and Europe. He had worked on North Slope developments for three years in BP's New York office and joined Alyeska in 1973 to help with the permit applications and the authorizing legislation in Washington. He had then gone with Ed Patton to Alyeska's interim headquarters in Bellevue—"my last job was to close it," Ron Merrett recalls—and was now manager of liaison with the pipeline's owner companies. In 1977, Merrett became manager of oil movements in Alyeska's operations division. Robert L. Miller had come to Alyeska as manager of public affairs in 1972 and saw a good deal of Washington until the pipeline bill was passed. From Nebraska, Miller had been an Anchorage newspaperman and then was press secretary for Alaskan governor Keith Miller.

On the second floor of Alyeska's offices, the engineering staff had been gathered in from Houston to be closer to the construction as it began. There were many new faces in the design management group, too—as well as many who had been on the project, either in Texas or Alaska, from the first days.

James F. McPhail, a civil engineer from Exxon USA who had worked on offshore platform and pipeline design in the Gulf of Mexico and Australia, became Alyeska's manager of engineering in 1973. Under McPhail was Alan A. Stramler as manager of field engineering and mile-by-mile design. Reporting to him was Dwayne R. Anderson, who had recruited himself into the project from C. F. Braun & Company as supervisor of design services. Exxon Pipeline's Joe Willing, already a project veteran, came up from Houston as manager of mechanical and electrical engineering for McPhail. On Willing's staff were John C. Wormeli as senior welding engineer, U. J. Baskert as hydraulics engineer, and C. Lawrence Johnson from Michael Baker, Jr., Inc. as mechanical engineer on mainline insulation and thermal heat pipe development.

Hal Peyton, now working for Atlantic Richfield on loan to the project, was manager of staff engineering under McPhail. The five groups reporting to Peyton were a mix of project veterans and newcomers. Robert J. Neukirchner, a young engineer from Exxon Research, came aboard in late 1973 as geotechnical supervisor. His senior geologist was Michael C. Metz from ARCO, who had been on the staff in Houston. Exxon USA's James A. Maple, also on the project since early in 1973, was stress analysis supervisor. Ralph D. Jackson, a SOHIO engineer with the project since 1971, was supervisor of river engineering. His deputy was Wim Veldman. ARCO's Al Condo, now a longtimer in Alaska, was supervisor of Arctic engineering. His senior agronomist was Joseph Neubauer from Exxon USA. Chris Whorton from ARCO, who had more time in grade on the project than almost anyone, was supervisor of engineering services. His senior thermal engineer was Christopher Heuer from Exxon Production Research in Houston. His seismic studies engineer was Douglas J. Nyman, who had just finished his doctorate in civil engineering at the University of Illinois, where two of his professors were Nathan Newmark and William Hall, developers of the pipeline's early seismic design criteria.

Geotechnical engineering and geology . . . civil and Arctic engineering . . . stress analysis . . . thermal engineering . . . seismic studies . . . river engineering . . . agronomics . . . hydraulics . . . mechanical and welding engineering. This wide-ranging catalog of engineering titles gave a pretty good clue to what kind of design project the pipeline had become. Over the years, there was a design progression that grew in complexity and sophistication as it went along—calling for just about every technical discipline in modern civil engineering.

Geotechnical engineering, as an example, is something of a new branch within the earth sciences field. Actually, explains Bob Neukirchner, it's a cross-discipline between geology and engineering. It deals with a range of things that crop up in civil projects: soil and rock mechanics, foundation design, the building of structures in or on top of the ground. The geotechnical work on the pipeline,

Neukirchner says, "has been the traditional earth sciences engineering that you'd encounter anywhere else . . . with, of course, the permafrost thrown in."

That's an important "of course," he emphasizes. Permafrost was apt to be found not only at any point on the route but also in great variety. "Frozen soils," Neukirchner says, "can be either elastoplastic or viscous. Some will degrade and reach an equilibrium . . . some will continue to creep for a long time. So you're dealing with a combined function of the type of soil, its density, its temperature . . . and time."

Stability of the soils along the pipeline has headed the list of geotechnical challenges on the project. By the time construction started, Alyeska and its contractors had already drilled several thousand bore holes along the route, analyzing several cores from each hole. Some of the borings were three feet in diameter, so that an engineer could climb down to see what the ground held in store for a pipelaying crew. The lesson was quickly learned that a "geologic surprise" might lurk just a few feet beyond any bore hole that was drilled, and even though the borings continued ahead of the ditching crews throughout construction, such surprises were appearing right up to the last mile of pipelaying.

Slope stability is important where the pipeline climbs or descends Alaska's mountains, and where it traverses or dissects the state's rugged hills—in other words, along much of the route. The design here had to assure against soil liquefaction and sliding. The more level river floodplains also had soils that might liquefy. Elevated construction was the frequent solution. "The support of the pipeline when it's above ground has probably been where we've spent the most time and effort," says Neukirchner. Then there were the earthquake fault zones. "Soils lose strength and become unstable because of earthquake motions," he says, "so you either avoid these areas, or design the system so that it won't be affected."

Seismic design began with the Menlo Park meetings between USGS scientists, the pipeline's engineers, and their consultants. The scientists' approach to the pipeline's earthquake risks was based on traditional seismology. But Doug Nyman points out that seismologists and engineers do not look at earthquakes in the same way. So each may take quite a different approach to designing a pipeline to withstand earthquakes.

"Seismologists put great store in readings of motions from instruments . . . using the Richter scale, which is also most familiar to the public," says Nyman. "Our fault zones on the pipeline are classified by the Richter scale. But Richter magnitudes mean little to design criteria. Engineers are more concerned with the *nature* of the motions. There can be a great difference in the effect of one large shock on a structure, or the effects of several repeated smaller shocks."

The pipeline's seismic design was a process of translating the science of earthquake monitoring into an engineering scheme. Before the pipeline was proposed, Alaska's only well-known fault zone was the Denali Fault, roughly 150 miles north of Valdez. Alyeska's own seismic studies identified three more zones north of Denali— the McGinnis Glacier Fault, the Donnelly Dome Fault, and the Clearwater Lake Fault. "We're not even sure today that those three zones cross our line. But we can't take the chance that they don't," says Doug Nyman. Alaska's worst earthquake, in 1964, had severely damaged Anchorage, and tidal action from a seaborne *tsunami* after the quake had hit hard at the old town of Valdez. The Denali Fault, though, had shown no displacement then.

Based on Alaska's earthquake history, the pipeline route was divided into seismic zones, each defined by a maximum Richter magnitude. From Prudhoe Bay to the southern foothills of the Brooks Range, the Richter maximum was 5.5—northern Alaska, in fact, had no real history of quakes. In the area of the Denali Fault and the other zones, the Richter maximum was pegged at 8.0. The area from Glennallen south to Valdez was given the highest rating: 8.5 on the Richter scale. In truth, says Nyman, "the entire pipeline system is probably designed to withstand an 8.5 Richter measurement."

Alyeska's seismic studies took in not only Alaska but recent major earthquakes elsewhere up through the early 1970s: the Tehachapi quake in 1952 and the San Fernando quake in 1971 in California, and the 1972 quake in Nicaragua. Industrial damage data from those events made one point quite clear. Damage occurred much less from earthquake motion than because poorly anchored equipment was torn loose and went banging around breaking itself and other equipment. That lesson is evident throughout the pipeline's seismic design. Everything on the line, including all equipment and the buildings that contain it, is qualified first to withstand shock—and then is anchored and braced so that nothing short of a seismic cataclysm several magnitudes off the Richter scale could jar it loose.

The pipeline itself is earthquake-proofed like no other oil transport line. The mainline is instrumented to detect shocks over its entire length around the clock and around the calendar. The instruments are not seismographs, which would pick up readings from a tremor anywhere in the north Pacific Basin. They are strong-motion accelerographs that react only to nearby movement. The units are at the terminal and, except for two in the north, at each of the pump stations, where they will pick up readings along the mainline between stations. They measure motion along a triple axis: vertical, and on two horizontal planes at 90° to each other.

Signals from the accelerographs go into a small computer at each instrument site, where they are processed and sent on to the main terminal computer. Within minutes after an earthquake is sensed, pipeline operators have a printed report of where it is, its magnitude, and its likely effects depending on whether or not it exceeded, and if so how far, the pipeline's design limits at that location. The report is in layman's language—"It will be data that anyone can understand," Nyman says. At the same time, the terminal will have been receiving its usual, constant readings of pipeline pressures all along the line, where any sudden changes trigger an alarm system. Based on all of this data, already partly analyzed for them, operators can decide if the damage has probably been severe enough to justify shutting down the system.

An earthquake need not damage the pipeline or even stop its operation. In early 1976 a severe earthquake ravaged northeastern Italy. At 6.5 on the Richter scale, it caused heavy loss of life and property damage. That area is the route of the Trans-Alpine Pipeline, a 40-inch oil line that runs about 300 miles from Trieste to southern Germany and has been operating since 1968. It was not designed against earthquakes. Its steel pipe has no special metallurgy. Yet the Trans-Alpine line took no damage during the Italian quake. It shut down only because two electric lines powering pump stations were cut when their poles fell.

By contrast, the Alaska pipeline's low-temperature steel gives it added strength. Its welds have been 100 percent X-rayed. It has standby power throughout the system. And in the earthquake fault areas, it is designed to counter quake movement. In the Denali zone, where the pipeline is buried, it is laid across a concrete track every sixty feet that allows it to slide twenty feet horizontally and also move five feet vertically. At one of the smaller fault zones, similar construction allows ten feet of horizontal shifting and seven feet of vertical movement. In the other two zones, the line is elevated and its support system also allows horizontal motion. So, when the earth moves, so does the pipeline. "Just like a strand of spaghetti," says Doug Nyman.

That's an apt description of the entire pipeline. A cross-country pipeline is a long, thin tube of steel. The Alaskan line is 800 miles long. Converted to foot-scale, its length is 4.25 million feet. Its diameter is just four feet. So it's about a million times longer than it is thick. And the pipe, of course, is hollow. Its walls are about half an inch thick.

Walk up to a piece of 48-inch pipe and hit it. It's pretty solid. But weld 800 miles of that pipe together and figure out some way to pick it up by one end. You could indeed wave it around like a strand of spaghetti. Now lay the pipeline. Fill it with hot crude oil. Half of the pipeline is buried in the ground, where the surrounding temperature will stay fairly constant. The other half is raised on supports, and the air around it will vary in temperature over a range of more than 150° from winter to summer. As solid as it feels, steel pipe "flexes." It expands and contracts with temperature changes in the oil within it and in the air around it. So a pipeline is not a static, rigid length of tubing. It moves under continual stress both from inside and outside. And this is especially so of a pipeline in Alaska's climate.

Stress analysis, says Alyeska's Jim Maple, has centered on the pipeline's above-ground design. But stress also occurs where the below-ground pipe bends to follow Alaska's careening topography and wherever a transition is made between the buried and elevated line. A buried pipeline is fairly restrained, although even its movement under stress will faintly deform the ground it's buried in. The elevated pipe's movement is not obvious to the naked eye. Yet it does move. Between the coldest and hottest extremes of its surrounding temperature, the line when empty could shift 12 feet in an 1,800-foot section. Even when fully insulated and filled with oil, it can move, from winter to summer, as much as two degrees of an arc.

By mid-1973 Alyeska was still refining its selection of construction modes along the route. Almost 360 miles of the line were to be conventionally buried or, where necessary, deep-buried. Some eight miles were set for refrigerated burial. About 50 miles, in river floodplain areas, were also still marked mostly for burial. And 380 miles were to be elevated on supports or pilings.

The pilings were unlike any that had been used in cross-country pipelining. In fact, they were no longer called pilings. They were "vertical support members"—or VSMs. A VSM was an upright stanchion of 18-inch steel pipe, and there would ultimately be 78,000 of them installed in pairs and spaced every fifty to seventy feet along the elevated line. Each pair of VSMs was connected by a steel saddle crossbeam, and the pipeline itself rested in a Teflon-coated shoe assembly that allowed it to slide laterally as the pipeline breathed.

"There was also the question," relates Jim Maple, "of what shape the pipeline should have as it goes across the ground. Stress forces are created by the cross-sectional metal area of the pipe . . . its diameter and wall thickness . . . and the temperature range, with the pipeline oil at 140° and the ambient air as low as minus 70°. The one other above-ground line, the 36-inch Trans-Arabian line, was as totally restrained as though it were below ground. We felt we could not design a restraint for these forces, so it evolved into an unrestrained or partially restrained line. We had to allow some movement. So, how do you allow that movement?"

Several ideas were looked at. Bridges also expand and contract and are built with small expansion gaps at each end to allow for this. A pipeline, welded from end to end, obviously could not contain gaps. Russian pipeliners had used a simple zigzag design, but this added greatly to a pipeline's length. The solution was a trapezoidal zigzag, permitting gentler offsets from the center line and also letting the line run straight again on the trapezoid's longer legs. To provide some necessary restraint, the pipeline is firmly anchored on VSM crossbeams without a shoe assembly every 800 to 1,800 feet at each end of the trapezoidal patterns and also at one point on the straight leg. Even these anchors, though, are on a metal slide that will allow movement in case of an earthquake.

If the elevated pipeline is not in contact with the permafrost, the VSMs are. Marginal permafrost, where the temperature may hover just around the freezing point, will react against an intrusion of anything slightly warmer than itself. The steel would partly thaw the permafrost, and two things might happen. The permafrost could reject the VSMs, "jacking" them back out of the ground. Or the thawing permafrost would allow the VSMs to settle, twisting the pipeline out of alignment. This is partly countered by corrugating the outside of the VSM, which adds to the bonding surface and moves the stress line out from the VSM itself, using the sheer-strength of the soil to work for the bond rather than against it. A sand slurry is also poured around and into the VSM, where it becomes like the permafrost itself and strengthens the bond.

The real answer to the problem, though, has been the thermal VSM. These were eventually used on almost 80 percent of the elevated line—the exceptions being in the far north where permafrost is thaw-stable, and in places where the VSMs rest on bedrock to begin with. Thermal VSMs are equipped with heat pipes, a nonmechanical, self-operating heat exchange system. Each VSM is equipped with a pair of tubes two inches in diameter, descending inside the VSM below the ground surface. Each tube is topped by an aluminum radiator, 10.9 inches in diameter, with twenty equally spaced fins. The tubes are filled with an anhydrous ammonia refrigerant that carries heat away from the permafrost up to the outside air through the radial fins. The refrigerant then condenses and returns down the tubes to begin the process again. This natural convection system keeps the permafrost frozen the year round.

The heat pipe research by Exxon Production Research and Woodward-Clyde was based on earlier work dating back into the late ninteenth century, says Michael Baker's Larry Johnson, the Alyeska project engineer on the development. The thermal units were manufactured by McDonnell Douglas Corporation in a converted aircraft paint hanger in Tulsa. The heat pipes also allowed VSM design lengths to be substantially shortened. Nonthermal pilings might have had to be sunk 150 feet to reach stable soil below the permafrost. With heat pipes, some VSMs are as short as 22 feet.

Larry Johnson was also the project engineer on mainline insulation. The above-ground pipe is insulated around its circumference with a 3¾-inch layer of Fiberglas covered with a galvanized steel jacket and clamped in place. This preserves the oil's warmth and keeps it in a pumpable state if the flow is stopped. "This is the most engineered insulation ever used," Johnson says. "We have cool-down and start-up problems here that had never been looked at before. The design specifies a 'worst-worst' case where the skin temperature of the pipe can never fall below minus 20° for a twenty-one-day period if there is a shutdown." The insulation was manufactured by Owens-Corning Fiberglas Corporation, and the contract included all of the panels, jacketing, and sealing accessories—including the mechanical manipulators that folded each 15-by-24-foot panel, more than 90,000 of them, around some 420 miles of pipe. Owens-Corning also made the reinforced Fiberglas modules filled with polyurethane that insulate the pipeline's shoe and anchor hardware where it passes over each VSM saddle.

The buried pipeline design, by contrast, was called conventional. The term was only relative. Burial depths ranged from three feet to more than twelve feet, depending on soil and terrain conditions. All pipe had been precoated with an epoxy material to protect against corrosion. All buried pipe was also wrapped with plastic tape. In addition, all underground pipe is laid on a bedding in which are buried two zinc-ribbon anodes that prevent chemical and electrolytic corrosion. Says Joe Willing, who helped design the corrosion system: "Zinc anodes had been used for shorter lines, but they are novel on a pipeline of this size."

At a few points on the route, neither conventional burial nor above-ground construction was possible. The pipeline had to be buried in permafrost, no matter what.

Three of these sections, totaling four miles in length, were in southern Alaska where the pipeline route crossed the Glenn Highway at Glennallen, and near the Gulkana River and Hogan Hill, where an elevated pipeline would have blocked a habitual caribou migration route. The solution was refrigerated burial. The pipe is buried, but it is first insulated like the above-ground pipe, and running along its bedding are two six-inch coolant pipes. Refrigerated brine is circulated through these lines at each location by electric motors in a nearby building, which also houses a heat exchanger and condenser to remove heat from the returning coolant and pass it to the outside air. The brine goes out to the coolant tubes at 8° and comes back at 18°. Refrigerated burial is also used in 200-foot pipe sections that connect two of the northern pump stations to the mainline.

Other special designs have been used in areas where it was feared that caribou and moose would not cross under the raised pipeline. In a few places, the line is built with an overbend arched high over the ground. But at twenty-two locations Alyeska engineers used a design they nicknamed the "dipsy doodle." The above-ground line dips below the ground for about a hundred feet, then reappears and resumes its way on supports. At these crossings in the north, the buried sections again are insulated and jacketed, and the burial ditches are lined with foam panels. In the south, in marginal permafrost, the insulated buried crossings are accompanied by free-standing thermal VSMs to aid in keeping the soil frozen.

The process of choosing between the below-ground and above-ground modes and all of their variations kept on into the construction years. Between mid-1974 and mid-1975, as the pipelaying was already starting, 167 more mode changes were made. The net result was a gradual increase in above-ground construction mileage. Each trapezoidal zigzag also added slightly to the overall length of the pipeline. In this way, the original 789 miles ultimately grew to 800 miles.

Throughout the design, the Alyeska engineers were aided by computers that were programmed to do everything from stress and seismic calculations to selecting hardware and even producing the detailed design drawings. The programs, moreover, could be chained together so that engineering data could be fed in one end and emerge at the other end as a set of drawings.

"The pipe test program at Berkeley essentially gave us our below-ground design," says Jim Maple. "The computer programs looked at stress analysis and seismic waves in a way that told you the combined stresses, and from these we arrived at the designs for the buried bends." Computers also worked out a set of standard configurations for above-ground portions of the line—such as the trapezoidal zigzags—and for the transitions where the pipeline went from one mode to another, entering or coming out of the ground.

A computer program of these configurations enabled Alyeska to compile what Maple calls "a cookbook for the detail design people . . . a 'Tinker Toy' set that was pieced together for the mile-by-mile design." Adds Maple: "Sometimes it seemed we had sold our soul to the devil. But that program helped us with our decisions, and we could still go to a hand-design on special problems where we had to patch in elements of several configurations."

Computers also helped in designing river crossings and flood-plain construction. The pipeline route lies within river floodplains for more than 50 of its 800 miles. It crosses eight hundred rivers and streams. "You have to take that with a grain of salt . . . we counted until we got to eight hundred and called it that," says SOHIO's Ralph Jackson, who was Alyeska's supervisor of river engineering. "In terms of scour, thirty-four of the streams were major—which means that at flood stage they had a potential of moving the stream bed to a depth in excess of five feet. There were eighty-four special designs on those streams, because we cross many of them several times."

Alaska's waterways are infinite in variety. There are "incised" streams that have one unchanging channel. There are streams with multiple channels, resulting in heavier scour where the channels come together. "Braided" streams have still more channels and subchannels, and are forever changing course over their broad floodplains—the Sagavanirktok River plain is almost a mile wide. Each crossing had to be designed for a maximum flood stage, arrived at by using a U. S. Army Corps of Engineers computer model that in some cases looked back over one thousand years of a river's history. Maximum scour depths were calculated by Dr. Thomas Blench of Northwest Hydraulics Ltd. in Edmonton, Canada, an expert on gravel bed stream behavior.

The country around the pipeline route also posed some challenging ground water conditions. Several glaciers in southern Alaska had a habit each year of damming up temporary lakes formed from their own spring and summer runoff. "On occasion," Jackson says, "those lakes will rise and burst the glacial dams . . . and all hell breaks loose." The nearby pipeline had to be built to withstand these possible annual glacial flash floods.

Then there was *aufeis*. This in German means "upward ice." Peculiar to northern regions, it is water that pushes its way up and out of the ground, fracturing the surface ice and forming new ice layers over it. It can grow to fifteen feet in thickness, migrating down slopes and over riverbanks as it forms. From end to end of the project, says Alyeska's Wim Veldman, "*Aufeis* imposed special design problems as well as a construction headache." In several areas, it threatened to cause flooding "right over the line," Veldman says. At one northern point, *aufeis* in 1975 was climbing to the top of a VSM.

All of Alaska's quirky natural phenomena had to be cranked into the river and floodplain design. If the pipeline was to be buried under a floodplain, Alyeska had two choices. It could use deep burial. "We have pipe buried twelve to fifteen feet below the scour line in some instances," Jackson says. Or it could use shallow burial and build river "training" structures to guide floodwaters away from the line. The structures commonly were "rip-rap"— dikes of boulders in graduated sizes up to three and a half feet in diameter and more than a ton in weight, some of them rising to ten feet in height and extending for several miles. Rip-rap was a technology in itself. Alyeska used standards developed variously by the Corps of Engineers, Colorado State University, and the California Divison of Highways.

Besides the Yukon crossing, more than a dozen other major river points were bridged. The larger bridges are free-span suspension structures. The span across the Tazlina River is 650 feet, and the Tanana River bridge spans 1,200 feet, with the pipeline at these locations slung in clamps or saddle-supported VSM shoes below a network of large-diameter braided cables that also braced it against wind. In 1975 it was decided to replace a planned buried crossing on frozen silts at the Gulkana River with a 400-foot tied-arch bridge.

Smaller stream crossings were made on steel plate-and-girder bridges, from 13 to 18 feet wide, built from standardized girder spans 180 feet long and used in up to three multiples. The girders were set on concrete piers, and in the piers were the 18-inch VSMs sunk into the bed of the stream and its banks and equipped with heat pipes if soil conditions called for it.

Government monitors insisted on wide safety margins in all of this design. An extra 20 percent in depth was to be added to the maximum scour level at buried crossings. But often, says Jackson, the rule was: "If a safety factor of 1.5 is good, you'd better use a factor of 2." In almost all cases, water crossings were built in winter when temporary disruption of the streambeds would not affect fish passage. If this was not possible, coffer dams were used to divert stream flows during construction. In any case, each stream after construction was to be left as it was found in terms of watercourses and velocities.

Up and down its 800 miles, the mainline is interrupted at intervals by massive chunks of finely engineered steel hardware: the pipeline's control valves. There are 151 in all, more than on any line of similar length in the world. All of them are made of low-temperature steel, and each weighs 70,000 pounds when oil is in the system. Most of them serve an environmental purpose.

The 71 gate valves are installed mostly in flat country and on downhill slopes, on both sides of river crossings, and in other sensitive locations. The gate valves are built to keep working at 70° below zero, and 62 of them are operated by battery-powered electric motors, controlled remotely by microwave and VHF radio from Valdez. The gate valves can be closed in four minutes to shut down or isolate the pipeline sections they control.

The other 80 valves, somewhat smaller in size, are check valves. They are one-way valves, each containing a hinged clapper that lets oil flow north to south in the pipeline—but not in reverse. The check valves are installed on uphill slopes and are meant to prevent large oil spills if there is an upstream break in the line that would cause the oil ahead of it to flow back down to the rupture point.

The big mainline valves are of a scale with the rest of the pipeline's hydraulics system. That system's main junctions are the pump stations spotted from the origin point at Prudhoe Bay to the final station at the Little Tonsina River about seventy miles north of the Valdez terminal. The size of all hardware on the project sprang directly from the size of the line itself. "The first decision was to buy 48-inch pipe. The design of everything else progressed from that," says Joe Willing. The next step, he says, was to pick the pipeline's prime movers—the engines to drive the centrifugal pumps at each station that keep the oil flowing against its own weight and over the ups and downs of the route. The engines are 13,500-horsepower modified aircraft-style gas turbines, and they were ordered from the Cooper-Bessemer division of Cooper Industries soon after the pipe in 1969. Ultimately there will be four at each station, three in use at all times and one as a standby.

Each of the pump stations itself has an intricate system of valves to control the flow of oil through it. One pair of valves, at the upstream and downstream ends of the station, lets operators isolate the station completely in an emergency. A second pair can divert oil into the station's 55,000-barrel storage and pressure relief tank. The third pair allows the oil to flow straight through the station, bypassing the pumps, during maintenance runs by mechanical scraper "pigs."

Siting the stations along the route was ruled purely by the pipeline's hydraulic needs over Alaska's terrain. The first four station sites are concentrated north of the continental divide to boost

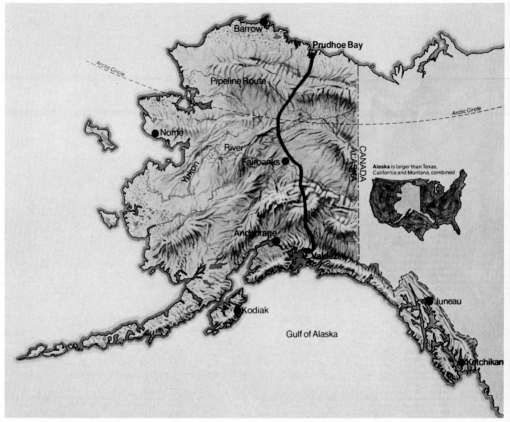

On its 800-mile journey from the oilfield on Alaska's North Slope
to the tanker terminal at Valdez, the Alaska pipeline crosses the state's
three major mountain ranges. It is the first oil pipeline in the world to
originate north of the Arctic Circle.

The pipeline route sweeps up from Prudhoe Bay to 4,790-foot Atigun Pass in the Brooks Range on Alaska's continental divide—the highest elevation the line must climb.

Galbraith Lake camp, on the north side of the Brooks Range, was one of 30 construction camps built as bases for the project. The camps were self-sufficient miniature towns, each housing up to several thousand pipeline workers.

More than 71,000 pipeline field welds were needed to string the line together from Prudhoe Bay to Valdez.

The first section of the pipeline was buried deep under the Tonsina River in southern Alaska in March 1975.

Around the calendar and around the clock, workers on the pipeline kept at their jobs in winter temperatures of 40° below zero and lower.

Because of permafrost and other ground conditions, 423 miles of the 800-mile pipeline are elevated.

More than 800 animal crossings are built into the pipeline to allow passage for the annual caribou and moose migrations across Alaska. Some crossings are buried "dipsy doodles" like this one. Others are high, arched overbends in the elevated pipe.

Alaskan bears, seldom intimidated by anything man does in their wilderness, were frequent visitors to project sites like Pump Station 5 in the southern foothills of the Brooks Range.

During and after construction, hydroseeding and other revegetation methods were used to restore the natural ground cover along the pipeline route.

Keystone's roller-coaster canyon edges were the scene for almost two years of some of the most difficult pipe laying on the project.

Permafrost and avalanche conditions at Atigun Pass in the Brooks Range called for insulating the pipeline and burying it in an eight-foot-square concrete box.

At the end of 1976, with years of engineering challenges conquered, some of the last pipe installed on the Alaska pipeline was laid in place over the high Atigun saddle.

Destination point: Valdez, on Prince William Sound in the Gulf of Alaska, is the site of the Alaska pipeline's 1,000-acre receiving terminal and tanker loading port.

the oil over Atigun Pass. The next four carry it downslope out of the Brooks Range and across the flatlands past Fairbanks. Two stations are on the uphill side of the Alaska Range below Isabel Pass. The last two lift the oil over the Chugach Mountains into Valdez.

Five of the stations on the upper half of the line are built on refrigerated gravel pads over permafrost, again using a brine circulation system under a plastic foam mat to keep the site stable. Turbines at the four most northern stations are run on natural gas brought from Prudhoe Bay in a separate 10- reducing to 8-inch pipeline. The eight stations farther south are powered by turbine fuel made in three small topping plants from oil bled out of the mainline at Stations 6, 8, and 10, and trucked to the other stations. All of the stations are totally enclosed against whatever the outside weather may be, and are manned by rotating crews of operators and maintenance workers.

In a pipeline design that breeds superlatives like Alaska itself, the destination point figures to be an engineering crescendo. That, in truth, is the pipeline's Valdez terminal. Like every other part of the Alyeska system, the massive shipping center was designed and engineered from scratch. At Valdez, especially, this meant literally from the ground up. Before anything could be built there, the sprawling thousand-acre terminal site had to be completely chiseled out of the rocky roots of the Chugach above Valdez Arm. The long task of site preparation brought together all that Alyeska had learned about rock mechanics, slope and soil stability, and the execution of yet another prodigious earth-moving exercise.

All of this was just to get ready to build the terminal's facilities. Three years later, these are arrayed on terraces planed into the mountainside. At the northwest corner of the terminal, the pipeline comes swooping in from its last descents through Thompson Pass and Keystone Canyon. The oil again is metered and identified, owner by owner. A system of manifold valves then routes it into storage tanks to await each company's tankers. The batteries of tanks have been placed on the uppermost terraces of the terminal site for two reasons. They are well above reach of seismic waves like the *tsunami* that rolled into Valdez in 1964. And they are designed to feed the stored oil down to the berthed tankers by gravity.

At full throughput, there may be thirty-two tanks in the terminal's two tanks farms. For start-up, all fourteen tanks in the east farm have been completed, and four are finished in the west tank farm. Each tank is 250 feet in diameter and 62 feet high and is topped by a rigid cone roof designed to stand up under the formidable snow loads in Valdez each winter. The tanks are paired off within containment dikes that will hold 110 percent of their combined oil volumes in case of leakage or rupture, plus another two feet of surface water that may have accumulated in the containment basins. Each tank has a capacity of 510,000 barrels. Together, the first eighteen tanks can store an eight-day throughput of North Slope oil at the pipeline's initial capacity.

The terminal's 37.5-megawatt power plant and vapor recovery facilities are also on the upper terraces. Both function as part of the environmental safeguards at Valdez. The vapor recovery system removes oil vapors from the air spaces at the tops of the storage tanks, keeping them from passing into the outside air. The power plant, besides generating all of the terminal's power needs, also injects inert gases from its boiler system into the tankage air spaces to provide a nonexplosive layer over the stored oil. The gases are sent into the tanks as oil is drawn off to load tankers and are withdrawn again as the tanks are refilled from the pipeline.

Downhill from these facilities, on the lower terraces, are the pipeline system's control center and ballast treatment facilities. Three 430,000-barrel ballast storage tanks, also cone-roofed and enclosed in containment dikes, receive the ballast piped in from the berthed ships. Ballast is allowed to settle in the tanks for at least six hours, and oil that comes to the surface is skimmed off to be cleaned and piped up to the crude storage tanks. The remaining ballast water is treated by chemicals and air flotation until it contains less than eight parts of oil per million parts of water. Then it is pumped into holding ponds on the terminal shore, from which it is sent as clean water into the sea through jet nozzles some two hundred feet below the surface of the bay.

Tanker berth design at Valdez is as modern as any present-day tanker port. There are now four berths. Three are built on pilings driven into bedrock, and each will accommodate 250,000-ton tankers. A fourth berth is floating where the bedrock sloped too steeply for pilings, but is anchored to shore by massive hinges. It will take 120,000-ton tankers. The piled berths are each equipped with four 16-inch articulated steel loading arms. The floating berth has four 12-inch loading arms. Ships can be loaded at a rate of up to 110,000 barrels an hour. In less than twenty-four hours, as a rule, a ship can arrive, berth, unload its ballast, refill with oil, and be on its way out into Prince William Sound again.

The tanker loading system is designed for maximum safety. Mooring lines are equipped with quick-release mooring hooks, operated from the berth and by remote control from the berth operator's control panel. In an emergency that would risk a moored vessel—or the shore facilities—the ship can be turned loose almost instantly. The oil loading arms have valves that can be closed in six or seven seconds from the birth panel and from the terminal control center, and also have quick-disconnect couplings, Alyeska's system-wide oil spill contingency plans provide for everything from oil dripped on the loading dock to larger spills in the waters of the bay.

The computer control center at Valdez is the nerves and brain of the entire pipeline system. The center has four computers—two to control pipeline operations, one primary and one standby with automatic failover, and two to monitor and record terminal operations, oil movements accounting, and seismic data analysis.

The computer console for the pipeline controllers displays every function on the line from Pump Station 1 at Prudhoe Bay to the point where the oil enters the terminal a few hundred yards from the control center. The control computer scans a total of about 2,000 status and alarm points on the pipeline and at the stations, taking a reading from each one every ten seconds. Operators can also issue control commands to more than 300 points throughout the system.

Similarly, the terminal's controller console displays the status of every mechanism within the terminal site, controlling everything at Valdez from the incoming oil to the tanker loading. Pertinent data from the upper recovery plant, ballast treatment systems, power plant, and other services are also displayed. Another computer system at Alyeska's offices in Anchorage received daily oil movements data and pipeline terminal volumes, which are continually matched against the arrival and departure schedules of the tankers and their capacities to carry the constantly flowing oil south out of Alaska.

The instant availability of information from any point on the 800-mile pipeline system depends on a triple-tiered communications system designed to assure against any break in the continuous data. The "backbone" communications network is a microwave system based at Valdez with forty relay stations spaced at line-of-sight distance of a mile and a half to forty miles apart along the route—one at each pump station, and twenty-eight remote repeater sites that are self-sufficient with their own fuel supplies and diesel power generators. Microwave is used for three forms of communication. One group of channels carries supervisory control and seismic instrument data. Other channels are for pipeline controller voice communications with the pump stations and with mobile radio-equipped vehicles and aircraft moving between stations on the lookout for leaks and other damage. A third set of channels is used for administrative and nonoperating telephone traffic.

The microwave network is backed up by satellite. Satcom II is in a stationary orbit 25,000 miles above the equator. Earth stations are at Pump Stations 1, 4, and 5, and at Valdez. If there is a break in the microwave network, all data and voice transmissions are automatically switched over to the satellite network. The third of the pipeline's communications systems is the remote gate valve control network. Each of the sixty-two remote gate valves is tied to the nearest pump station north of it and then on to Valdez by microwave and also by an independent VHF radio channel. The remote gate value supervisory system at Pump Stations 3 through 12 is designed to interrogate each remote gate valve at one-hour intervals for a check on its operating condition. A status change or alarm at a gate valve will be reported immediately by the valve supervisory system to the upstream pump station, which then alarms the pipeline supervisory system for display to the controllers at Valdez.

As the pipeline's design came together in 1973 and 1974, so did Alyeska's organization to manage the building of it. There was some shifting around at first until the construction supervision fit comfortably. Within Peter DeMay's project management group, construction was divided into two main parts. The pipeline and roads were to be under one manager, with an assistant for each type of work. The pump stations, terminal, and communications were put under another manager, again with an assistant for each area.

That group was set first. Kenneth E. Anderson came to

Alyeska from Mobil Oil in 1972 as project manager. By mid-1973 his three deputies were in place: Rodney B. Higgins from SOHIO for the pump stations, Vic D. Filimon from SOHIO for the terminal, and William J. Benton from ARCO for communications. Hollie Childress now went back to Exxon Pipeline after five years on the project, the last six months of it with the Fluor and Alyeska design group in Los Angeles. Lionel B. Morrow of Exxon took up that assignment.

Thomas J. Kofodimos of Atlantic Richfield was named project manager for pipeline and roads in mid-1973. The pipeline and road work, it was decided, should be combined geographically, with one manager for the work north of the Yukon and another for the work in the south. Nathan E. Bauer from Exxon took the assignment for the northern area. The southern area manager was a newcomer to Alyeska, a 39-year-old civil engineer from Chicago named Frank P. Moolin, Jr.

11.
A Different Project
Now...

Clearly, the design of the Alaska pipeline by 1974 had gained a sophistication and complexity that beggared the feasibility and engineering studies first put together in Houston five years earlier. This was quite a different pipeline now from the $900-million project that TAPS, early in 1969, had first proposed to build across Alaska. It was also quite an advance over the more complete design that had evolved by the time of Alyeska's first project description in mid-1971. And the design was still progressing. It would keep doing so even as the crews started laying the first pipe sections in 1975.

Why? Was the project in its earliest years in some ways underengineered? Without the delays of the native land claims and environmental lawsuits, would a pipeline based on those initial plans have been built—and failed as disastrously as all of its opponents were predicting? Did the delays give the pipeliners the chance to catch up with the engineering technology they badly needed to build a proper oil pipeline out of the Arctic? And allow the government to catch up as well on its watchdog role of protecting the Alaskan environment from a botched-up early design?

"No,"said Hal Peyton.*"To sit back on the outside and say that government . . . or 'we the people' . . . is the only thing that kept the industry from raping Alaska, I think, is a statement made from extreme ignorance," Peyton maintained.

Peyton spoke as a civil and structural engineer with long experience in Alaska—and with long experience in oil company operations there. "I have lived in Alaska for all of my professional life . . . and on the North Slope for four years in the late 1950's," he told the Department of Interior hearings in 1971. "I have been continuously engaged in either research, development, design, or teaching related to permafrost problems for the past seventeen years, fourteen of which were with the University of Alaska."

As a consultant in the project's earliest days and then as Alyeska's manager of staff engineering, Peyton's involvement in the pipeline was total, and of an intellectual depth that few could match. So was his expertise in every aspect of the project. As a consulting engineer at the start, he had a central role in setting the technical framework on which the design was built. He gave expert testimony in the pipeline's congressional hearings from 1969 on. He helped write the project descriptions and environmental statements. He helped on legal briefs when Alyeska was allowed to intervene in the environmental lawsuits. He was in Houston during the first design years and helped build the engineering staff there and in Alaska. Later he worked on procuring equipment and mobilizing the construction effort. Then he was back on the design staff in Anchorage.

His own succession of tasks, Peyton noted, pointed up that the Alaska pipeline has been built in a new way. It's been a first-of-a-kind project in every sense—not the least of which is that the work has gone on under a number of new and important constraints. Some of these were technical and environmental—"trying to build a project in the far north, with its technical and construction problems," he said, "and the complexity of trying to convey to people, unaccustomed to that environment, how to design and construct there."

Other constraints, Peyton said, came from the new climate of government intervention and regulation—"the passage of the National Environmental Policy Act in the middle of our project,

*Hal Peyton died unexpectedly in March 1977 just a few months before completion of the Alaska pipeline.

A Different Project Now . . .

which caused a radical change in federal intervention, and in private industry activities as a result." Another factor was what Peyton called "the utter, sheer size of this project . . . until you've lived with it a long time, you really don't begin to comprehend how massive it is." Part of this, he also noted, was the size and awkwardness of the rapidly growing organization that Alyeska had to set up to carry out the project.

Before the Alaska pipeline, this was not the way big oil industry projects got built. "When oil companies went about a project," Peyton explained, "they characteristically operated with a great deal of speed and a great deal of aggressiveness in the sense of getting on with the job and getting it finished."

Peyton cited an example of a project from his own earlier experience during the industry's years in southern Alaska—long before anything like a National Environmental Policy Act was on the lawbooks. "This was the first drilling platform built here in the Cook Inlet," he related. "It was a very large project in its day. It was also . . . like this pipeline . . . the use of extended technology. And it was built in a very harsh environment, due to the high tides and high winds and extremely high ice forces in the Cook Inlet in wintertime."

That project, like the pipeline, also moved quickly at the start. "In less than a month," Peyton said, "the research work to develop the design conditions was started, the design itself was started, the steel was ordered, and some of the equipment was ordered . . . all at once. And we sat down and developed a strategy . . . what the risk of having something go wrong was, and how much contingency did we need in the steel ordering, the plate sizes and all that stuff, and basically what the final design was going to look like . . . we did all of that."

But the Cook Inlet was hiding a problem. "We worked on that platform design for a year, and lots and lots of design was complete, and the steel was coming out of the mills," Peyton went on. "And at that point in time, our research work came up with the answer that the whole design concept relative to the ice loading was wrong. Instead of the ice putting a steady load against the platform, it turned out that it was highly dynamic. The ice was going to just shake the hell out of that platform."

This, of course, was not anticipated. "But it happened, and it was serious, and it controlled the design from then on," said Peyton. "Because of that problem, we had to redesign the whole platform. But the platform was put in place, with the new design, with the right steel, on time, and in budget."

That was in 1962. "Okay," said Peyton, "you could then say . . . what if somebody had questioned us at the time of the original design criteria of that platform, and pulled us up short at that time, and said: 'You guys don't know what you're doing . . . that design is all screwed up . . . it's a good thing we caught you up short or you'd have made a bad, disastrous mistake.' Well, in that instance, that wasn't true. The first design concept was wrong. But that design was never built."

The Alaska pipeline, Peyton pointed out, was begun in the same way. "The research and development work was started, and the design was started, in mid-1968. By early 1969, the steel was ordered, and a design was put together for the purpose of initial construction planning, for the procurement of materials and machines and camps . . . and for the purpose of acquiring the necessary permits."

Here, though, the parallels between the two projects stopped. "Characteristically, in the past, permits had never been a problem on a pipeline," Peyton said. "Pipeliners were accustomed to very sketchy designs that said: 'This is what we're going to do, and now we want a permit to go across the land to do it.' But it was then that the inquiry of this project was started, by individuals and organizations, and bureaus and permitting agencies. That very preliminary design was questioned in a highly technical manner, and out of context, and it was found to be not satisfactory."

The same would have been so of the early Cook Inlet design—or, probably, of the preliminary design on any big pipeline project in the past, if it had been so closely scrutinized at the very start. The Cook Inlet engineers found their own mistakes, corrected them, and went ahead to build a successful platform on schedule. And the same thing was already happening within TAPS. "In 1969, prior to any strong government intervention in this project," said Peyton, "I filed a report as a consultant to the project. It said, in effect: 'We must change the design because the design, as it currently is, is guaranteed to fail.' And that report resulted in a radical change in the way we were going about the project."

The important point was that the oil companies had no intention, at any time, of going ahead to build a big, expensive mistake. They expected, as they had always done in the past, to keep refining the Alaskan design before the construction phase started—the

A Different Project Now . . .

first year had to be spent building the road, anyway—and to correct their construction problems in the field as they went along. For this reason, the preliminary pipeline plan was a dead issue, within TAPS, before it ever became a public controversy outside.

"Now, you can ask two questions," Peyton wound up. "One, would that early pipeline have been a good pipeline, had it been built? And the answer is . . . No, that would not have been a good pipeline. I don't know how to answer whether it would have done the job or not . . . as a minimum, we'd have had a helluva lot of maintenance on it. But you must also ask . . . Would that early pipeline have been built without government intervention? My answer to that is that, with or without government intervention, it would *not* have been built."

By 1975 a pipeline was being built. Its design, however sophisticated it had become, would go on adapting to the realities of building in Alaska all through the construction years. Meanwhile, Alaska itself was adapting to the realities of the pipeline project.

12.
...And a Different Alaska

On the drive in from the airport, the Chugach Mountains behind Anchorage stood out sharply in the clear, mild 34° weather. It was January 1975.

"Up here to look for a job on the pipeline?" asked the bearded young cab driver. "People are really flooding in . . . I must haul six or eight a day from the airport. One guy last week cracked me up. I asked him if he was going up to the Slope to work, and he said: 'What's the Slope?' Then I told him guys were getting eleven hundred dollars a week up there driving a rig and his ears kind of pricked up."

Alaska had seen it all before. For eighty years or so—first by way of its southerly seaports, then over its new roads and railways, and in recent times through its gateway airports—Alaska had braced itself against one in-pouring after another of the "Boomers"—the seekers after whatever fortunes were said to be up for grabs at the moment. Now it was happening again. With construction moving into high gear this year on the Alaska pipeline, Alaska was booming once more. This time, though, it was a boom with a difference. Now that it was finally under way, it might go on for decades. Its dimensions figured to be superlative even by Alaskan standards.

Already, the impacts of the pipeline project were being felt across every part of the state's life, from native villages to urban areas, from the man in the street in Fairbanks and Anchorage and Valdez to the officials in the state government in Juneau—and beyond to Seattle and the Pacific Northwest. With thoughtful planning, the long-range benefits of what was getting under way today might still be multiplying for Alaskans well past the year 2000, as the state's vast natural wealth is converted into a broad material affluence and—hopefully—a new social order for Alaska's scattered but fast-growing people.

Alaskans were inured by now to "future shock"—both its promises and its pitfalls. They had seen, angrily at times, what had been done in the past to their treasured way of life in the name of progress. Some felt that this was only more of the same. "Most of us old-timers who came to Alaska twenty or twenty-five years ago are appalled at what's going on today," said Claire O. Banks, executive vice-president of the Greater Anchorage Chamber of Commerce. "I raised three kids here, and we used to go everywhere . . . now there are twenty guys at every one of our fishing holes."

Kenneth C. Hume, a financial consultant who was president of the Anchorage chamber in 1975, felt that "it's not a matter of stopping the pipeline . . . it's just being able to control the size of the wave, of keeping it from becoming a tidal wave." Hume then was also head of the Downtown Anchorage Association and executive director of the Alaska Bankers Association. From this multiple vantage point on Anchorage and the state as a whole, he said: "Whatever way you look at it, the economy here is going to continue to grow."

Too often, Alaskans felt, they had come up empty-handed after earlier periods of intensive exploiting of the state's resources. They were determined this time to see that Alaska itself got a bigger piece of the action as it unfolded—that the benefits and profits did not all flow away with the North Slope's oil as it was shipped Outside to the Lower 48 states.

"This time," said Frank H. Murkowski, president of Alaska National Bank of the North in Fairbanks, "we have to list our priorities. We have environmental concerns here because, frankly, Alaska has a long history of being raped. But Alaskans are basically conservationists . . . not preservationists. Today we have to balance the drawbacks in what's happening against the potentials."

It had been a rainy weekend this January when thousands of prominent Alaskans slogged through ramshackle downtown

Juneau, tucked into the southeastern panhandle, to inaugurate Jay S. Hammond as their new governor. There were dress balls at the Baranof Hotel, the Elks Club, and the National Guard Armory. There were receptions at the city's new Hilton Hotel and aboard the ferryboat *Columbia*.

Jay Hammond, a stocky, trim-bearded man in his early fifties, looked like an outdoorsman even in a business suit. For these events he was in tuxedo. So was the new lieutenant governor, former state senator Lowell Thomas, Jr., whose name in national news stories invariably was followed by "son of the newscaster." The wives—pert Bella Hammond and stylish Tay Thomas—were in evening gowns. So was the Hammonds' 15-year-old daughter Dana. Older daughter Heidi, 17, arrived on the *Columbia* in a parka. "She was not alone," said a press account, noting the mix of mink coats, muumuus, and work clothes among the well-wishers.

Jay Hammond was governor—but not by much. He had edged into office by all of 287 votes among the 91,393 total votes cast in the election the previous fall, and it had taken more than three weeks of recounts to decide who really did win—Hammond, the Republican candidate, or William Egan, the Democratic incumbent governor. Since statehood in 1959, Alaska had only had four governors. On his way to Juneau, Hammond had beaten out all three of his predecessors: Republicans Walter Hickel and Keith Miller in the primary, Egan in the main event.

Of all of the candidates, Hammond had been the darkest of dark horses. He thought he had finished his state service—six years in the Alaska House of Representatives and six more in the state Senate, the last two as Senate president. In 1972, gerrymandered by Egan into what loomed as a losing race for Senate reelection, Hammond "retired" from politics. Born in Troy, New York, Hammond was studying petroleum engineering at Penn State when he went into World War II as a Marine fighter pilot. He came to Alaska in 1946, took a degree in biology at the University of Alaska, and went to work as a pilot-agent for the U.S. Fish and Wildlife Service. He also met and married Bella Gardiner, a native Alaskan, her mother an Eskimo and her father a Scot. The Hammonds moved to Naknek on Bristol Bay at the shank of the Alaska Peninsula and started an air taxi and guide service, which they intended to go back to operating now.

But Jay Hammond was talked into making the Republican race in 1974, and now he was governor. He had campaigned as a conservationist on a theme of planned, controlled growth for Alaska.

... **And a Different Alaska** 133

Sitting at his desk in Juneau's musty capitol building, Hammond also talked about balances in the state's life.

Trade-offs, he conceded, were inevitable. "We can't preserve Alaska as we've known it. We're going to lose certain freedoms and qualities of life here," Hammond was saying. But, he added: "People talk about 'balanced development.' I want *imbalance* . . . in that the benefits outweigh the costs. We are programmed along certain inexorable courses . . . we're going to have growth, and the impact of that growth, no matter what we do. The idea is to channel that growth, and learn to contain it. It may be a lesser of evils . . . not making things better, but keeping them from being as bad as they might be."

The new administration's director of policy planning and development was Robert B. Weeden, a biologist and land use expert from the University of Alaska's Institute of Social, Economic, and Government Research in Fairbanks. Weeden's research had been studied during the early field work on the pipeline. "We do," he said now, "have the power to manage this growth, so that when we get 600,000 people in Alaska, it isn't a total disaster. We still have a chance to prevent what's happened Outside."

Alaska's story for the foreseeable future, as Hammond was suggesting, would be one of impacts—demographic, economic, social, environmental, civic, and ethnic. Alyeska itself had sponsored the first socio-economic study of the project's possible impacts in 1970, and also had helped fund an early planning study for the city of Valdez.

Several studies added to this data. In a 1975 pipeline impact study for the U.S. Department of Labor by Human Resources Planning Institute in Seattle, Alaska's population was projected to increase to 481,600 by 1980—almost 50 percent higher than its 1975 level, and four times its level after World War II. The state's work force, said Theodore Lane, the institute's director, would grow about 20 percent from 1975 to 1980. Both employment and unemployment rates would stay proportionately as they were. There would be more jobs, but also more joblessness. By 1990, said another study by David T. Kresge at ISEGR in Fairbanks, Alaskans' per capita income would rise to $4,500, from its present level of about $3,100.

Each of these factors would ripple out over Alaska's economy in varied ways. There would be impacts on skill levels, salary struc-

134   . . . And a Different Alaska

tures, local prices, housing costs, and real estate values. Victor Fischer, ISEGR's director, was worried about the effects on Alaskans who had to live on fixed incomes. Impacts were also to be expected on what economists call a state's infrastructure: its transportation and communications systems, its educational and health care facilities, its banks and other businesses. Growth would affect the costs of government itself, at both the state and local levels—more people needing services and an expanding economy calling for a greater variety of services.

A whole range of impacts would be translated down to the community level. Rising in-migration and business activity would put pressure on local utilities, schools, roads and traffic, law enforcement, and—even in boundless Alaska—on air quality. And, of vital importance in Alaskan life, there would be the social, cultural, and ethnic impacts. The effects of an accelerating growth economy would converge on a largely rural, frontier life-style, and particularly on the large Alaskan native population just starting to share at last in the state's wealth. "We already have the nation's highest unemployment rate . . . coupled with a village poverty scale which points up by contrast the affluence of Appalachia," said Governor Hammond in his first state budget message. In large areas of the Alaskan economy, commented H. Lee Atherton, vice-president in charge of the Alaska Department at Seattle-First National Bank: "Social and economic impacts are a paradox . . . what's economic for us is social for them."

The tone of Alaska's future was more cheerful long range than short range. History and geography had conspired to give Alaska a unique economic profile. "Alaska," said Robert R. Richards, who is now executive vice president of Alaska Pacific Bank in Anchorage, "is an utterly fabulous economics laboratory." This meant taking the bad with the good, Richards said, pointing out that "last year, for the first time in at least ten years, the rate of inflation in Alaska exceeded that of the United States nationally." For all of its natural wealth, Alaska still had to import much of what its people needed, from basic food and clothing to major durable goods. The cost of living for Alaskans ranged up to almost half again as high as for the rest of the U.S. Climate and transport costs only partly explained this. "The small size of Alaskan markets limits competition and minimizes the gains usual in larger scale economies," explained University of Alaska economists Arlon R. Tussing and Monica E. Thomas.

High incomes and a growth economy—but also, high unemployment, living costs, and inflation. The Alaskan fabric was woven of paradoxes. A major reason, put simply, was that Alaska—physically—was a big state where big things were apt to happen. But Alaska, in terms of its economy and people, was small. Superimpose a massive project like the Alaska pipline on a modest economic base—the cost of the pipeline was already a many-times multiple of Alaska's current gross state product—and there would be problems.

The state was scrambling to catch up. But like an economic fault line down its length, the construction of the pipeline was sending tremors through Alaska that would take time to subside. They were being felt most harshly at first in and around Fairbanks—"Hub of the Trans-Alaska Pipeline," as the *News-Miner* put it. Fairbanks had sought out its role as the center for construction activity. But Fairbanksans had geared up for the action once before, when the oil was discovered in 1968, and some of them had got caught out badly by the four-year delay. They were being more cautious this time. "We did have a boom in 1968, and that's the reason I don't think we're going to have one now," said Dr. William R. Wood, the retired University of Alaska president who was now executive vice-president of Fairbanks Industrial Development Corporation. "Fairbanks has matured enough to avoid the boom-and-bust cycles."

Wallis C. Droz, general manager of the Municipal Utility System, had been city manager of Fairbanks for ten years. He wasn't so sure. "I think it's just starting to happen. This is going to go on for another ten years," he said. Would it change Fairbanks? "No," said Droz. "There are a lot of individualists here, and there always will be. Fairbanks has always been kind of a melting pot anyway . . . the 'Last Frontier' . . . and that character will not change."

Frontier or not, the pressures on city systems in Fairbanks in 1975 were intense. The city had already expanded its water and sewage facilities five years before. It was now realigning its cramped streets for better traffic flows and, hopefully, less air pollution from the burgeoning vehicle traffic. But telephone service in Fairbanks, for both local and long-distance calls, was badly overloaded. And because of manufacturers' backlogs in the Lower 48, the city couldn't get delivery on new switching equipment for another two years.

Most of the strains in Fairbanks were "people" problems. The area's population of about 55,000 was pegged to grow to 78,000 by

1980. "But we didn't have a large population here in the first place to absorb all this," said Joe E. La Rocca, a newspaper reporter who was then director of the Pipeline Impact Information Center for the Fairbanks North Star Borough. La Rocca felt that the city was in for a long surge of activity, with ups and downs along the way. Meanwhile, he and his staff were meticulously monitoring the social landscape from housing and supermarket prices to drunkenness, drug use, prostitution, and even the local divorce rate. Fairbanks had its own "DEW Line" on the problems of growth.

The Anchorage area, with triple the population of Fairbanks already, was expected to grow 50 percent in size by 1980. Being larger, Anchorage was perhaps better cushioned to absorb pipeline impacts. But it was feeling the early shocks too. "We've got big-city problems we never used to have," said Claire Banks at the Chamber of Commerce. "We're choked. Our traffic now is so congested . . . I can hit any stop street and I'll be eight or nine cars back. We don't like that." Kenneth Hume was worried about the first full year of pipeline construction ahead. "When spring arrives, the barges will be going to the North Slope, and they'll overwhelm us . . . just passing through, they'll drain the community's resources." Hotel space in Anchorage was jamming up already with mounting volumes of pipeline business travelers.

Alaska's banks were booming as the state's economy heated up. "As a result of the pipeline delay, Anchorage is better prepared than it would have been three or four years ago. But there's a great need for additional financial institutions in the state," said Dick Fischer. an Anchorage real estate developer who was the chief organizer of a new bank. Short-run deposit volumes were one thing, though, and long-term profitability was another. "Banking costs here are going up higher than in other parts of the U.S., and unless the deposits we get are long term and stable, we are going to have a profit squeeze," said Rodney W. Burgh, executive vice-president of Peoples Bank & Trust Company in Anchorage and the 1975 president of the Alaska Bankers Association.

Nonetheless, Alaska's twelve established banks had just topped an aggregate $1 billion in deposits for the first time, and they were about to be joined by three newcomers. One was a state bank, United Bank Alaska, chartered by five of the twelve native corporations created by the Alaska Native Claims Settlement Act of

1971. "Natives look at this as a way to get a good rate of return on their funds, and to have access to other sources of financial and investment counseling," said Britton E. Crosley, a Cook Inlet native who was a CPA with Arthur Young & Company in Anchorage and was helping to organize the native bank.

Hot-eyed bankers and brokers from all over the United States had rushed up to Alaska after ANCSA was passed with advice on how the natives should put their $962.5 million to work. The natives weren't having much of this. The regional corporations, and now the new bank, were bringing experienced management from Outside into Alaska. But they firmly intended to control their own affairs. "They want to show that they can invest their own money and be successful at it," said Crosley.

Native job opportunities were also improving. Early in 1975, Alyeska and the Department of Interior arrived at the terms of the Affirmative Action program required by the pipeline permit. Alyeska pledged to offer up to 3,500 job openings to Alaskan natives by means of institutional and on-the-job training programs— and actually hired more than 5,500 natives. It was also offering at least $220 million in contract bidding opportunities to native and other minority companies. About $130 million in contracts had already been let. And the pipeline company had set up Alyeska Investment Company, a minority enterprise small business investment company, with a pool of $1 million for loans and other assistance to minority businesses in the state.

"Historically, the industrialization of Alaska has never touched the Alaskan native. This time, it has," said Thomas G. Evans, a civil engineer and Doyon native. Evans was technical assistance program director for the Alaska Federation of Natives and, later in 1975, would be appointed deputy commissioner of the state Department of Labor in Anchorage. "It's important to be getting the natives trained, because there's no doubt going to be more pipelines up here, and now we'll have an industrial labor force that's ready for industry to draw from," Evans said. "Alyeska and some of its contractors are giving us an opportunity for our management people too. So some of our people can come back from other parts of the U.S. . . . there are jobs for them now."

Even in Alaska's climate of high unemployment, Alaskan natives statewide at times were 60 percent unemployed, and that figure went much higher in remote areas. The pipeline was the premium job market for everyone in Alaska now, as well as for those from high joblessness areas in the Lower 48. The job competition could be rough—and expensive. In Tommy's Elbow Room on Second Avenue in Fairbanks, job-seekers this winter were trading drinks and rumors. "A few months ago, three hundred dollars to a union business agent would get you on a hiring list," said a discouraged young native hunched over his bottle of Olympia beer at the bar. "Now it's up to fifteen hundred."

All through Alaska, in 1975, there were these contrasts between a disjointed present and an affluent future. The United States government had already authorized more than $60 million in funds to improve city services and transport systems in Alaska. The money was going to expand utilities, housing, school and health facilities, roadways, and airports—all focal points of pressure from the pipeline activity. The state was parceling out $10 million in Oil Development Impact Grants to five burdened locales. In Valdez alone, where the population had already grown from 1,143 to 3,100 in two years, a grant of $2.9 million was being used to build classrooms and to expand fire and police services. Other federal, state and private industry funds were being spent on manpower development programs for Alaskan residents, with emphasis on the natives.

In Juneau the state's projections were that, at 1975 world prices, Alaska's income from North Slope royalties, severance taxes, and pipeline taxes would top $1 billion per year by 1980. The University of Alaska's David Kresge estimated that this would rise to $1.6 billion annually between 1983 and 1990. But the state's oil income would be based on the difference between the price of the oil at the wellhead and the costs of moving it to market. With the pipeline's costs escalating, a squeeze was developing on Alaska's future royalties.

Meanwhile, the state was still on the financial brink. "Without added income," said Lawrence C. Eppenbach, deputy commissioner for the Treasury Division of Alaska's Department of Revenue, "we'll be out of liquid cash by the spring of 1976." To span a $350-million to $400-million funding shortfall until the pipeline came onstream in 1977, the Hammond administration planned to brake state spending and tighten up on existing corporate taxes. A possible state lease sale in the Beaufort Sea east of the North Slope was considered and then postponed. "The state can take a more objective view about everything now," said George W. Rogers, an economist with ISEGR in Juneau. "It has more options, more room to maneuver."

State planners, using a computer model, were casting up scenarios on how best to combine Alaska's options to get through the short term crunch. One of these, as Larry Eppenbach put it, was the "Doomsday" option—a tax on oil and gas reserves in place, before production started. Alaska had not intended to use it unless there was a new delay in completing the pipeline. Now with the postponement of the Beaufort Sea sale, the tax was put through in 1975.

But not without a stiff fight between Republican Governor Jay Hammond, the Democrat-controlled state legislature, and all of Alaska's divergent interest groups. Like the long pipeline debate, and the election of Hammond himself, the reserves tax again revealed Alaska as a state split right down the middle—its frontier psyche in torment over what shape the future should take. It was the conservationists against the boomers all over again. Odd coalitions kept appearing. Hammond wanted the tax because he felt Alaska should support itself by ongoing revenue-producing measures and not by selling off more and more pieces of its resource base. Tax supporters in the legislature felt the oil industry simply wasn't paying its share of the freight. Arrayed against the tax were the oil companies who felt the pipeline costs and ad valorem taxes were enough to bear at this point. And they were joined now by many of the natives, whose corporations were making exploration deals with oil producers on ANCSA lands, and who feared the tax on reserves would put a damper on the search for new oil in Alaska.

It was a perfect portrait of Alaska's ongoing dilemma. Alaskans welcomed development for all of the economic benefits it could bring to them. But they also spurned it for all of the changes it might bring to their life-style. "It depends upon to whom you're talking," said Hammond. "Those who are intimately plugged into the pipeline construction program, and profiting handsomely, think oil is wonderful. Those who are not feel it has eroded their quality of life. So half of Alaska concludes that oil is the best thing that ever happened to them. And the other half concludes that it's the worst."

The debate would go on. Some Alaskans also wanted to open up the state's other resources to development now that the pipeline had breeched Alaska's northern interior. But if there was any sort of fragile consensus in the state in the mid-1970s, it was that whatever was done, it should be planned carefully first.

"We're always saying how much better we do things in Alaska than those idiots Outside who have messed up their yards," said Jay Hammond early in 1975. "But we really haven't done things better. My concern is that if we focus entirely on resource extraction in years to come, someday, when those tubes go dry, we'd better still have some timber on the slopes and some fish in the seas." It would be some time, he knew, before the oil fields and Alaska's other resources went dry. "But," said Hammond, "that's been our problem in Alaska. We've been less than concerned enough about the day *after* tomorrow."

13. 1975...Toward the Halfway Mark

The floodplain of the Tonsina River is lightly wooded land that rolls up toward the northern foothills of the Chugach Mountains. About seventy-five miles above Valdez, the river meets the Richardson Highway north to Fairbanks. On March 27, 1975, with brief ceremonies, the first 1,900 feet of the Alaska pipeline were buried beneath the Tonsina and its banks.

The Tonsina was ditched first by rugged tractor-backhoes to depths of eighteen feet below the stream bed and eight feet below the maximum scour depth of the river channel and the floodplain on each side of it. Then the welded pipe sections were laid out beside the ditch. The pipe lengths in the 300-foot section to go under the river itself had been precoated with nine inches of concrete to combat the buoyancy of the empty line—each 40 feet of pipe weighed 80,000 pounds. A long row of twelve tractors with side-mounted lifting booms picked up this section in webbed slings and held it while the other sections to complete the 1,900-foot burial were welded to each end. All three of the long sections had been hydrostatically pressure-tested before they were laid out. After all of the pipe was lowered into the ditch, the sections leading back from each bank were weighted down with dozens of nine-ton concrete saddle blocks. Bulldozers then filled the ditch in with gravel over the pipe, restoring the river bottom and banks to their normal contours.

A few miles south, the Tonsina branches into the Little Tonsina River. In coming weeks, two more buried crossings totaling 2,100 feet of pipe were made under the Little Tonsina. In the same area, before the end of April, the first 1,800 feet of elevated pipeline were installed.

The pipeline was on its way. Progress goals for 1975 were ambitious. This was the first full construction season on the pipeline system itself. By year-end Alyeska hoped to have 45 percent of the mainline finished. The 800-mile pipelaying job was programmed to come along fastest of the system's three major parts. This included clearing the right-of-way, laying down a gravel work pad to protect the land from the constant traffic of heavy construction equipment, and—in this first year—installing almost 400 miles of the pipe. Buried portions, including river crossings, would come along most rapidly. The elevated portions, which included several river crossings over bridges yet to be finished, were the more intricate parts of the line, and could be expected to lag the burial work.

Pump station and terminal work was also scheduled to lag the mainline in terms of percentage completion as the project went along. That didn't mean the station and terminal contractors and their crews weren't starting off as intensively. Work had begun, in fact, at Valdez and at each station site during the past winter, before any pipelaying was possible. It would accelerate now as the weather improved, and would continue through the next winter, and each winter, when the pipe crews had to shut down. Site preparation for the terminal and stations was more complicated than ditching or elevating the mainline, and would take up time during these early months of 1975 before any of the terminal and pump facility structures could begin to rise. Already, though, concrete ringwalls for the storage and ballast tanks at Valdez were being poured, and drilling for the marine berth trestle piers was under way.

By midspring, nearly 12,000 workers were already spread from end to end of the project. The work force would peak at more than 21,000 by late summer. Employment on the project was being built up carefully, and most of the union jobs had to be secured before a worker even got to Alaska. Alyeska, the trade unions, and

1975 . . . Toward the Halfway Mark

the state of Alaska were putting out warnings by every way that the word could be passed that no one should just head north in hopes of getting some kind of work. But the job-seekers were flocking in anyway.

It wasn't hard to see why. Construction jobs throughout the recession-slowed Lower 48 were hard to come by. The Alaska pipeline, with a mounting visibility in nationwide newspaper and magazine stories, was a beacon for unemployed thousands all over the country. Unscrupulous employment agencies from as far away as New York were sending up applicants with promises of jobs that didn't yet exist and that they had no say in filling. By air and ship, and over the Alaska Highway through Canada, the prospectors for pipeline jobs came pouring up in Alaska like the Klondike gold hunters almost eighty years earlier. Once there, often without funds to live on, a worker might wait months to get to the top of a hiring list. Alaskan officials even considered setting up stations at the state's entry points to turn back anyone who couldn't prove he had a job waiting.

As the construction pace built up, the years of engineering and design work came up hard against the realities of putting a pipeline over Alaskan terrain. With the crews at work, the detailed mile-by-mile design was now confronting what actually lay under the land. What Alyeska had known for some time was coming true. Adaptation—both of design and of construction methods—would be a continual process from now right through 1977.

Frank Moolin had become senior project manager in charge of the pipeline department at Fort Wainwright outside Fairbanks. The sandy-haired civil engineer was already an experienced supervisor on refinery projects in Singapore and Europe, on the design of a U.S. Atomic Energy Commission nuclear waste storage facility, and on BART—the Bay Area Rapid Transit system that links San Francisco and its surrounding cities. He also had put in a year of work toward an MBA degree in business management.

The Alaska pipeline, said Moolin, "is a civil engineering project that happens to have a pipeline associated with it. Subsurface conditions are the factor here. If you're building something like the World Trade Center in New York, your exposure to subsurface work is early and brief. Here we'll encounter it to the last foot on the last day. On this project, the design is never frozen."

Alyeska's construction management organization was also

adapting almost from the start. Some management responsibilities had been delegated during the rush to get going when the permits were granted and the haul road was pushed through. The main example was Bechtel's role as construction management contractor for the pipeline and roads. Alyeska now found that changed conditions called for a different role for Bechtel. Moolin urged his boss, Peter DeMay, to draw some of Bechtel's duties back into Alyeska and put the pipeline department in direct charge of supervising the execution contractors. So, in May, Bechtel was shifted to become construction technical services contractor, supplying support services such as quality control in the field to Alyeska and the other contractors.

"We are not cutting Bechtel off as a key management arm. We are just reducing its length," explained Ed Patton. "With all of the planning work essentially done, we were in a sort of steady state operation, rather than a growth operation," added Peter DeMay. "We had a one-on-one relationship where we had too many of our people talking to too many of Bechtel's people. Bechtel people were between us and the execution contractors, and we came to the conclusion that we should wipe out one of those layers."

There were other factors. Bechtel, some felt, had brought too many people to the project, but not enough of the right people. Basically, though, there was a clash in philosophies on how to build the pipeline. Bechtel, with its reputation for designing and building big projects all over the world, wanted a "turn-key" contract on this one—total responsibility from start to finish. Alyeska's owner companies saw that this wouldn't work. Alyeska had designed the pipeline, not Bechtel. Already there were arguments about why the design had to be built so strictly to specifications. Alyeska, from its years of studies, knew exactly why each design element was in there. Nor was Bechtel doing the building. The execution contractors were. Alyeska was the common thread through the entire project.

The situation with Fluor Alaska was different. On the stations and terminal, Fluor was doing much of the design work in tandem with Alyeska. Fluor was also directly supervising the subcontractors at each site, rather than execution contractors who were spread all over each long section of the mainline route. Fluor's top management was also keeping in close monthly touch with the project. Around the world, Fluor had often performed turn-key contracts too. But not on this project, and so it did not impose an extra management level between Alyeska and the work.

With construction now in high gear, the Alyeska owner companies set up a more formal management system of their own to keep watch over the expensive undertaking. Joseph D. Harnett, executive vice president of SOHIO in 1975 and today the company's president and chief operating officer, took over as chairman of Alyeska's construction committee. At the same time, within that committee, a smaller *ad hoc* committee was formed by the project's major owners—SOHIO, BP, ARCO, and Exxon—to parcel out some of the specialized advisory tasks such as technical, legal, employee relations, and public affairs counsel.

Joe Harnett, a New Jerseyan with more than thirty-five years in engineering and management jobs at SOHIO, has likened the construction committee to a corporation's management committee. "It gave the owners a focal point for their inputs into Alyeska, and it had the role of communicating to the owner companies the needs and problems and expectations of the project," Harnett explained. "Alyeska as a non-profit company was a cost-allocating organization. The construction committee has been a basis for budgeting, and for providing the information the companies have needed for their financing purposes. The cost estimates generated by Alyeska have been reviewed and approved by the construction committee, with the owner companies having the final word on major matters."

The *ad hoc* committee apportioned other tasks among the owner companies by means of a number of advisory subcommitttees. The work in each case was supervised by one of the major companies. Exxon coordinated technical and marine matters. ARCO looked after the areas of administration, accounting, auditing, and tax advice. BP's specific concerns were employee and public relations. "That made it possible," Joe Harnett pointed out, "for me at SOHIO to coordinate the overall project work." Also, he noted: "We were also working out a permanent operating agreement between all of the owner companies for the time when Alyeska would phase over from construction to the operation of the pipeline system."

More immediately, in 1975, the construction committee had just reviewed the project's still-rising costs. In June Alyeska issued a new estimate: $6.4 billion. The cost was up almost five billion dollars from 1969. It could be estimated that more than three billion dollars of that was the result of inflation, particularly in materials costs. Environmental expenses were probably adding another two billion dollars. With two full years of work yet to go, the company said it couldn't be certain about a final price tag. Labor productivity in Alaska's climate was bound to be a constant variable. So was the availability of specific skills. So were deliveries of materials and equipment. The work was speeding up to meet the mid-1977 operating deadline, and to avoid the costs of an extra year of construction. Design adaptations might add to costs. So might unexpected and especially difficult pieces of construction—sure to crop up under actual field conditions on a project of this size.

One of these was already cropping up this summer: Keystone Canyon.

Between Alaska's gold and petroleum bonanzas had come its copper boom. Keystone Canyon, just north of Valdez, was a gateway to the rich ore fields in the Copper River Basin. In 1906 a New York-based Alaska Syndicate, which included Daniel and Isaac Guggenheim and J. P. Morgan, was blasting a railway through the narrow scenic divide of Keystone Canyon. Then the financiers shifted their railhead first to Katalla and next to Cordova, where a rail route to the Copper River could also bring out coal deposits along the way. In 1907 a Valdez copper developer tried to take over the Keystone route, and the climax came one September morning in a bloody gunfight between the crews of the New York syndicate and the local promoter. The partly tunneled railway through Keystone was never finished.

Now, in mid-1975, Keystone Canyon echoed again with the thunder of another construction saga. No gun-packing railworkers this time, but a battalion of frustrated men and machines. And the scene wasn't the canyon floor, but the crest of its high east wall. Alyeska and its Section One contractor, MK-River, couldn't build in the canyon because the pipeline would preempt the only highway route north from Valdez. The alternative was the canyon lip. This part of the route was hardly four miles long. But it gave the pipeliners more problems than any one hundred miles somewhere else. The first problem in 1975 was getting construction equipment up the steep faces at the north and south ends of Keystone to the working sites near the top of the wall. Only tracked vehicles— tractors and bulldozers—could clamber up those slopes, dragging materials and other equipment up behind them.

Once there, though, the rock faces were found to be too dangerous to drill. The work had to be moved above the rock. Now Keystone Canyon's logistics took to the air. Over a long weekend in August, helicopters hoisted almost 250 tons of heavy construction machinery up to four landing points on the bench above the canyon. The workhorse of the operation was a Sikorsky S-64 Skycrane that could carry ten tons in a single load. Four smaller choppers carried the lighter loads. The bigger pieces of equipment were taken apart and flown up in pieces to be reassembled at the top. Smaller machines like compressors were taken up intact. By September, having saved weeks of time with the airlift, the work crews were going all out again, working in from each end of Keystone and also north and south from the four canyon-top staging areas.

Up north, Alaska's contrary polar weather was staging another drama this summer. The short period of warm months each year was the only time the North Slope could get supplies by sea. Each summer, the 455-foot Bureau of Indian Affairs ship *North Star* brought a year's worth of food, fuel, and other goods around to Barrow from southern Alaska. Now, with the Prudhoe Bay oil field under development, the brief retreat of the polar ice pack each July and August was also the signal for an annual armada of barges, towed by tugs, to bring up huge tonnages of oil well supplies and field production equipment. The shipments in 1975 totaled 168,000 tons. Large parts of this were prefabricated building modules—some of them weighing 1,300 tons and standing nine stories high—that had to be installed at the North Slope in time to start feeding oil into the line in 1977. Some pipeline materials were also on the barges this year—most importantly, the small-diameter pipe for the gas fuel line to the first four pump stations.

From as far away as Houston and Japan, the fleet of forty-seven barges and twenty-three tugs rendezvoused at Seattle early in July. By August they had passed Icy Cape in the Chukchi Sea; confident of a good head start on the receding ice pack. But the ice wasn't receding this year. Curiously, North Slope natives had warned this spring that the pack might not move—for the first time since 1897. All through August and into September; the barges could get no farther than Wainwright, still almost three hundred miles from Prudhoe Bay. The U.S. Coast Guard icebreaker *Glacier* was disabled trying to push a lead through the ice. Around the U.S.—and at the scene—oil company managers, bankers and stockbrokers, and journalists kept an anxious daily score on the struggle.

More icebreakers and other ships were brought up to help. Ten barges made it through in early September, despite hull damage to six of them and some of the tugs. One barge was beached in stormy weather. Another nineteen, carrying cargo not needed so urgently, turned back to Seward and Anchorage, where their cargoes would be sent north again by rail and truck. The last seventeen barges waited it out until October 5, when they were finally able to run into Prudhoe Bay and stay there, stuck fast a mile and a half out in the frozen shallows of the Arctic Ocean, to be off-loaded during the next winter. The *North Star*, at the same time, got to within a mile of Barrow, where the coming year's supplies were helicoptered in to the town's 3,000 people.

The 1975 barge episode was a new chapter in Alaska's age-old story of Arctic challenges. Failure to get the shipments through would have had only a small effect on pipeline construction. But if the production modules and supplies had not made it, a serious dent would have been put in the oil field's own tight time schedule. Another massive shipment was due to come up in the summer of 1976. The word was only half-jokingly passed around the oil companies that, before the fleet started out, it would be wise to ask the Eskimos this time what the ice pack was likely to do.

Rough terrain and unpredictable weather were to be expected in Alaska. Meanwhile, this summer, something more ominous had surfaced on the pipeline project.

For possibly the first time on a long-distance pipeline, 100 percent of the girth welds around the pipe's circumference were being inspected by X ray. Alyeska had subcontracted this work north and south of the Yukon to two companies, to be monitored by Bechtel's quality control staff as a support service. In July Alyeska found discrepancies in radiographic records in Section Two that required thirty-nine welds to be X-rayed again, and seven others to be cut out and replaced with new welds because the X rays could not be confirmed. In August, more X-ray problems were found in Section Three. A sample audit by Alyeska raised the possibility that some X-ray records were being falsified.

Ketchbaw Industries was the subcontractor for all radiography south of the Yukon, including Sections Two and Three. In September one of its former employees, Peter Kelley, filed suit against Ketchbaw, charging that he had been fired because he wouldn't collaborate in covering up falsified X rays. Kelley agreed to help Alyeska identify the false records—the first step toward finding out if bad welds were being covered up with X rays from good welds.

The problem, it seemed, was that the subcontractor was cutting corners to get its record-keeping done on schedule. The result was that Alyeska now had to question its records on all welding done so far on the project. By the end of 1975, that would come to some 30,800 welds. Alyeska began a full-scale review of all radiographs. Welds that could not be confirmed as acceptable were X-rayed again or repaired.

At the end of the year, Ketchbaw's contract was terminated and a new subcontractor was named. Alyeska itself now took on the responsibility for interpreting all welding radiography. Alyeska itself had first uncovered the problem. But the damage had been done. Congress and the regulating agencies were voicing doubts about the pipeline's quality control, and a controversy over welding records and repairs began that would take more than another year to resolve.

On the Yukon River, late in the summer of 1975, the twin hovercraft *Yukon Princess I* and *Yukon Princess II* glided ceaselessly back and forth through the gathering dusk of an Alaskan September afternoon. Sometimes they raced—for candy bars as prizes.

High in the control cab of *Princess I*, mechanic-welder Andy Rogers of Associated-Green, the execution contractor for Section Four, said he was winding up his year's work. Since breakup last spring, for twelve hours a day, seven days a week, with only a week off here and there, Rogers had been piloting the hovercraft as they ferried giant rigs loaded with 80-foot lengths of double-jointed pipe, and smaller trucks with other supplies—115 vehicles a day, all bound north up the line across the broad, rambunctious Yukon.

Since January, at something like $1,700 a week, Rogers had collected gross pay of about $53,000. So he was taking the rest of the year off. "The government will just get it all from now on anyway," said the young pilot, scanning the laden deck below and punching the buttons that operated *Princess I* as his "ship" neared the landing incline on the river's north bank.

At Five Mile Camp, a short way above the river, the kitchen

crew was routing an Alaskan brown bear poking hopefully around the back door. The bear knew what it was doing. The food at the pipeline's twenty-nine construction camps had got such a reputation that the kitchens vied with each other to turn out meals of a remarkability this wilderness had never exactly been noted for. Harlan Elsasser, the affable resident camp manager at Five Mile in 1975, claimed his workers ate better than anyone on the line.

In the camp's two long dining halls today, Five Mile's off-shift workers were stoking up on a midday buffet spread that would have done credit to a Houston hotel. Out on the nearby pipeline work pad, a welding crew was seeing to its own lunch: thick steaks with all the trimmings cooked over roaring grills they had flanged up out of scrap materials. Do-it-yourself meals like this were against camp rules. But as pipeline projects around the world had come to know, the men from Pipeliners Local Union 798 out of Tulsa, Oklahoma, liked to set their own rules.

Hovercraft races on the Yukon and welders' steak fries on the line were ways of lightening the pressures of a construction effort going at top speed. The working pace along the pipeline was severe this fall. Up and down the work pad and at the stations and terminal, time was starting to run short on the 1975 building season. In the rest of the U.S. September mostly meant late summer or, at worst, the first faint chill of oncoming autumn. In Alaska, September meant winter was on the doorsill.

The Arctic ice pack wasn't the only place where winter was bearing down early this year. At Valdez, work on the tanker terminal was also besieged by weather of a sort that only Alaskans could understand or tolerate. Valdez was the terminal site because it could promise ice-free port conditions the year round. But winter could still start early there. As residents of the Alaska south-central coast knew from years of experience, the area could easily get 360 inches of mixed rain and snow between the end of summer and the next spring.

This September, the rains came. A routine day's trip in and out of Valdez often turned into a stay of several days while stoically patient passengers sat, smoked, and slept in Polar Airway's cramped old airfield shack. The heavy downpour was snarling the project's work for miles around. Out along the pipeline's Section One north of Valdez, four inches of rain in a two-day period washed out parts of the compacted gravel work pad. Rock and mud slides hit the Richardson Highway, route of the "Sourdough Express" rigs that were still carrying double-jointed pipe lengths out to the line from the wrapping and staging yards at Valdez. Sections of the access road into the terminal were also washed out.

The rains hit hardest within the terminal site. One massive slide brought down 12,000 cubic yards of rock above the vapor recovery area. Crews set to work to reterrace the hillsides around the terminal. Where possible, more of the unstable rock was carved away to form new benches in the slopes. Large faces of the remaining rock were literally bolted to the mountain—hundreds of holes were drilled horizontally, and then 30-foot rock bolts were inserted and grouted in place.

All of this meant more work for everyone from the construction crews back to the engineers who had to design new solutions for each problem that came up. The wear and tear of the first full construction season were obvious at every level of the project. Hal Peyton, in a rumpled sport shirt, was summing up engineering developments for an Anchorage visitor one Saturday evening before flying off early Sunday morning on some field site visits. "I'm sure you've seen it . . . our people are terribly busy, and they're awfully tired," said Peyton.

Ed Patton was just back from a Construction Committee meeting in the Lower 48. "Our problems right now are almost the *force majeure* kind of thing," he said with a glance at the rain sloshing down outside his office windows. Weather was one kind of force. Labor unions, all year, had been another. They couldn't strike the project. But they could find ways to trip it up and slow it down. "Safety" meetings and jurisdictional disputes had been chronic all during 1975. The afternoon before, coming in from the aiport, Patton had stopped off at the Anchorage Westward Hotel to address a regional conference of Allied Daily Newspapers editors and publishers. With his usual candor, Patton said of the labor problems: "It's a 'king of the hill' contest. The welders consider themselves to be the Marine Corps of the national labor movement. Teamsters Local 959 thinks it runs the state of Alaska . . . and that's about right."

Even with the Project Agreement, labor-management relations could be sticky. Alyeska had to deal directly with each of the unions. But it was also caught in frequent crossfires, particularly between the teamsters and the welders. The real and alleged issues ranged from safety to racism. At the Tonsina construction camp, an isolated racial scrap between teamsters and welders on a bus in

August had grown by September like an untended ditch in the tundra—a deepening crevasse of bitterness that was threatening a project-wide work stoppage. Because of the ironclad labor pact, this never happened. But through the fall months, wildcat walkouts and threats of more delays were still flaring along the length of the project.

Hot-eyed reporters were finding the breathless stuff of muckraking journalism in all of this. "Blood, toil, tears and oil" was the title of an article in *The New York Times Magazine* in July. Writing about safety conditions on the pipeline project, its author quoted speculative estimates that up to five hundred people would be killed in accidents before construction was finished. Alyeska replied that the project's injuries and fatalities so far were well below the mortality tables for industrial activity of any kind—less than half the historical rate for heavy construction projects in particular. The state of Alaska, it said, had projected 45.3 deaths for the man-hours worked during the project's first year—the actual number of deaths had been 12. By the end of the project's second year, Alyeska could say later, injury rates on the pipeline were still about 28 percent below the national average for similar projects—and the fatality rate was less than a fourth of the state's projections.

The Los Angeles *Times* sent a whole team of reporters up to probe the project. "Alaska Today—Runaway Crime and Union Violence" was the opening headline on a two-day, seven-article series in November. The pipeline, said the newspaper, was spawning a "crime wave" in Alaska. The teamsters had a "stranglehold" on the state. The *Times* cited a U.S. Department of Justice memo estimating that theft and fraudulent billing on the project might total as much as $1 billion. Pounding away at the theme into January, when Alyeska raised its cost estimate to $7 billion, the *Times* noted that costs were up $1 billion from late 1974. "Every penny of that massive overrun," it editorialized, "will come out of the pockets of consumers through higher prices for the oil that eventually will flow from the southern end of the 798-mile pipeline"

Not only was the newspaper out of date on the length of the pipeline, but it also displayed some ignorance as to just how oil was competitively priced in world markets. The billion-dollar theft charge drew an angry blast from Ed Patton. "If someone stole all the pipe and all of our construction camps, the total take would still be considerably less than a billion dollars," Patton said. Added Frank Moolin, who made the rounds of the construction sites almost weekly from Fairbanks: "I haven't noticed any camps missing."

There was also a report by columnist Jack Anderson in November that ten miles of 48-inch pipe had been stolen from Prudhoe Bay. This would have been more than 300 truckloads of the specially-sized 60-foot pipe sections, used at the northern end of the project. It also required special trucks to carry it. "We'll be surprised if the line winds up ten miles short when it's finished," commented Alyeska public affairs manager Bob Miller. "And surely someone would have seen it going down the Alaska Highway."

Much of this overblown journalism was bracketing the project's real troubles. Alyeska did have theft and wastage of some severity, but far below what might be expected—less than $1 million on a $7 billion project—and many stolen items were recovered. At the Glennallen construction camp, $14,000 in tires and other stolen items were recovered. Seven men were arrested, six of them workers on the pipeline. Pilferage of small tools and spare parts was also a chronic headache.

The problem was that in the rush to mobilize and get off to a fast start, Alyeska hadn't yet perfected its inventory and cost controls. Whatever was needed was often reordered to get on with the job. In some cases, tools had been incompletely numbered, so it was hard to know what belonged to whom. With a high turnover of new workers, security was loose over what anyone was lugging home in a duffle bag on "R&R" trips or what was being mailed out in packages in the meantime.

How much was theft actually adding up to? Again, the construction industry had actuarial tables for pilferage on large projects. This could run up to 4 percent of the labor cost on a reasonably paced project, and up to 7 percent on a project in a hurry— which, for better or worse, the Alaska pipeline was. By these yardsticks, theft on the pipeline this year might have been anywhere from $40 million on the low side to $70 million on the high side. *Newsweek*, in a report about the Los Angeles *Times* series, quoted another west coast reporter as saying: "They took the normal corruption of a $6-billion project and made a big deal about it."

But even at less than $1 million, that's still a lot of trucks, tires, and tools. Alyeska began tightening up its controls. Its new manager of security, as of October, was Robert J. Sundberg, who had just spent five years as chief of police in rough-and-tumble Fairbanks. He was based right in the center of the project at Fort Wainwright. Construction equipment was being fully inventoried. Scattered warehouses were being consolidated, and their contents logged. Tools were numbered and recorded as they were issued. Packages leaving the camps now had to be wrapped with a security guard looking on. Drivers who left running vehicles unwatched were ticketed the first time, then relieved of the vehicles—and therefore their jobs—after a second offense.

As winter closed down over Alaska in 1975, Andy Rogers was long gone from the Yukon. After September, he had headed Outside . . . "Mexico," he said, "and maybe Amsterdam. But I'll be back at the end of the year, probably on one of the southern sections." The *Yukon Princess*es were obsolete now. The steel bridge over the river was finished in October. From now on, it would carry all of the surface traffic to and from Alaska's north. And next summer, slung along one side, the Alaska pipeline itself.

Alyeska was adding up how far the pipeline had gotten in 1975. The 400-mile goal for the mainline had fallen short. By mid-December, when weather forced the pipelaying crews off the line, 371 miles of pipeline were installed. This included 208 miles of buried line and 148 miles of above-ground line, plus another 15 miles of buried or raised line at river crossings. Another 50 miles of pipe were welded and ready to be installed. About 90 percent of the work pad was finished. Another 225 miles of the right-of-way had been ditched for pipeline burial. A total of 46,800 of the eventual 78,000 VSMs were in place. All told, 56 percent of the mainline work—right-of-way, workpad, installed pipe, and VSMs—was complete.

The pump station and terminal work was more than 25 percent finished. This would keep going during the winter, when a good many of the tasks could be done indoors or in partly sheathed structures. The project workforce had peaked at 21,600 in September. It was down to less than 8,000 now. The terminal crews were holding steady at about 3,100 workers, awaiting delivery of critical hardware and materials to continue building at full strength. There were 1,800 workers at the pump station sites, and more might be added over the winter.

In Fairbanks the pipeline department was looking back over a hectic first year. Frank Moolin, behind his desk at Fort Wainwright, was finding it hard to relax during the enforced winter layover. The project had been visited in late November by President Gerald Ford and Secretary of State Henry Kissinger, on their way through Alaska bound for Peking. Moolin was their tour guide. Newsphotos showed him where he liked to be: out on the line, his trademark Irish tweed walking hat mashed down on his head while everyone else—including the President of the United States—wore the mandatory hard hats.

"We've learned an awful lot," said Moolin of 1975. "This year was a pull-ourselves-up-by-our-bootstraps operation. We didn't really have our organization set until July, and it wasn't working at top effectiveness until October. We got into a lot of bad habits in '75 because of this. The falsification of X rays by the contract radiographers was a debacle . . . so there was a germ of truth in the L.A. *Times* series. We've recovered from some serious setbacks, not always smelling like a rose . . . but standing on our feet."

14.
1976...Toward Completion

On a snow-crusted wall outside the workers' dormitory at the Alaska pipeline's half-built Pump Station 1, scant miles from the frozen surface of the Arctic Ocean, was carved the season's farthest-north graffito.

"Life Has No Purpose," it said.

It could seem so toward the end of an eight-week work stint on this frigid coastal plain in January. Yet from the few hours of half-twilight through each long night, the scene at Prudhoe Bay this winter belied pessimism. Across the white sweep of the North Slope for miles in all directions winked purposeful lights of Arctic oil activity. The temperatures here this month were hanging around 45° below zero—a few degrees higher at midday, a few degrees lower at night. That was in still air. With a few miles per hour of wind, equivalent temperatures were reaching minus 60° or 70°—or worse.

No matter. Strung out along the miles of ice-hard gravel spine road that wound through the Prudhoe Bay oil field was a hustling community of some 8,000 people. Newcomers were arriving at Deadhorse Airport as often as five times a day on Wien Air Alaska's scheduled jets from Fairbanks and Anchorage. Dozens of smaller corporate and private craft came in daily at airstrips hard by the Atlantic Richfield/Exxon base camp a few miles to the east and the British Petroleum/SOHIO base camp a few miles to the west. Between these points, company signboards looming out of the ice fog would read comfortably to an oil worker anywhere in the world: Halliburton, Dowell, Camco, Ralph M. Parsons, Banister-Stewart, Parker, Rowan, and many others as familiar in any oil patch.

Around the calendar and around the clock, in weather that fell off the bottom of the thermometer in winter and could go into the 80s in summer, the work of developing Prudhoe Bay's oil went on. So did the work on the pipeline to take the oil south. Roughly centered between the spreading networks of wells on each side of the oil field was the site of Pump Station 1. The ten-acre unit would be the most complicated of all of the stations, for it was where the first North Slope oil would enter the pipeline system for the start of its initial journey to Valdez a year and a half from now. Out on the dim flatness beyond the south edge of P.S. 1, you could see the 48-inch open end of the above-ground pipe, waiting to be hooked into the station later this year.

Winter had shut down all of the pipelaying along the right-of-way. But it scarcely dented pump station activity. Rod Higgins was project manager for station construction at Alyeska's offices in the Polaris Hotel in downtown Fairbanks. "We haven't had a wintertime slowdown," Higgins said in mid-January. "We've been active at all stations, getting the buildings to where they can hold heat and doing the piping and electrical work that can be done now. We're also doing some work on the gas line that will supply fuel to the first four northern stations. What you'll see at Pump Station 1 now is indoor work. At least, I *hope* that's mostly what you'll see."

Visitors to P.S. 1 the next day did see plenty of indoor work. And outdoor work—on the gas line, on the storage tank containment area, and everywhere else that heavily clothed construction crews could go, 45° below zero or not. Against the backdrop of a night world made workable with portable high-intensity construction lights, face-masked figures in bulky parkas and Arctic boots moved between the steel-clad buildings and the tank site. Running motors of trucks, bulldozers, rock saws, and heavy cranes billowed the air with clouds of condensing exhaust. Fueled on the run, the machinery operated for days at a time before heading into the temporary air-inflated "beluga" structure that was the site's maintenance shop.

Bill Windecker often didn't bother with a truck. At 7 P.M., when the thermometer outside his office was going north of minus 50° again and the brief twilight had long since given way to blackness—the sun hadn't come over the horizon here since late November—he set off to conduct a walking tour of the station.

Windecker was Alyeska's project manager at P.S. 1. He'd been hired for his talent at seeing that things got built right. He had done this for Aramco in the Middle East and for ITT in Alaska, and on other assignments from Canada and Greenland to Korea and most of Europe. The bluff New Jerseyan, whose site at the head of the pipeline drew visitors of every stripe from congressmen and the press to Alyeska's owner-company brass, would have been just as successful as a diplomat in the same Third World locales where he had gotten things built. Explaining the challenges of the North Slope, its inhabitants, and their work was no trickier than running a project in Iran or Korea. Windecker loved every minute of it.

P.S. 1 was Bill Windecker's personal kingdom in these years. He pushed the work along firmly but encouragingly, spending the long days in constant motion around the site, talking to everyone, missing nothing. The camp had a population of 270 workers this January, including 23 women, one of whom, Chrissie Storey, was the resident camp manager. At peak, later this year, there would be 430 people here. Fluor Alaska was in direct charge of the work. "I don't deal with the crafts . . . Fluor does," Windecker explained. "I monitor Fluor."

Windecker could tick off P.S. 1's details without breaking stride between the station's dozen or so structures. "This is 4,800 p.s.i. concrete. The spec was 3,000 p.s.i. . . . for whatever that's worth," he said in his usual way of describing things, stomping on the floor of the main pump building to show that it was more than half again as strong as it probably needed to be. "This is 1,200 cubic yards of concrete, two feet thick with reinforcing bar. We did it in one consecutive pour . . . nineteen hours and thirty-six minutes . . . that was last summer, not in this weather." And he headed off through the enclosed walkway to the booster pump building. His guests—trying to take photographs and notes, dodge electrical lines and the work crews, and keep their feet, fingers, and noses from snapping off in the wrenching cold—were hard put to keep up with his pace.

For all of this, and despite the early bad weather last fall, this winter had been reasonably kind to the work at the stations and terminal. On the North Slope, even with the low temperatures, the air had stayed relatively quiet. Winds could go up to fifty miles an hour or better in the Arctic, producing chill factors as low as 120° below zero—at which point men and machines have long since been forced out of action. Deliveries had kept up, so that the stations and terminal had gone ahead at high levels as hoped. These parts of the pipeline system were making up great chunks of their gaps in percentage completion over the winter.

"This is a system. If one part of it doesn't work, nothing else is going to do you much good," said Rod Higgins in Fairbanks. "Early in the project, the emphasis was on the pipeline portion . . . bed space was even limited for station people. Until we said: 'What's the good of having that tube if you can't move oil . . . the purpose of the whole thing is to move oil.'"

The stations and terminal also needed a high degree of "special item" equipment. The piping had to be made with a special low-temperature metallurgy, and the welding of it had a special chemistry. So did the rest of the system. "Many of these specifications have been used for other projects," pointed out Ken Anderson, senior project manager for the terminal, stations, and communications in Anchorage. "But we have stress and thermal factors, and snow loads, and seismic anchoring requirements . . . it's the combination of these things that make the design more complex and this pipeline different. And the big difference is the stipulations that were issued and the monitoring of these stipulations by federal and state agencies." The result was a great many long-lead-time components. "Damned little of this equipment was a standard catalog item," said Rod Higgins. "Some of it that was ordered two years ago is just coming from the manufacturers now. Our pipe fittings are special metallurgical castings. Our gas turbines have a control system on which some additional engineering has had to be done. And turbines are always a long-term delivery item anyway."

Out at Fort Wainwright, the pipeline department was gearing up again. The goals for this second year of mainline construction were as ambitious as—or more so than—the first. "We're expecting to have all of the line installed, insulated, and hydro-tested by November 1," Frank Moolin told his staff in January. There were, admittedly, some tough construction hurdles to get over in 1976. Dwayne Anderson, supervisor of design services in the mile-by-

1976 . . . Toward Completion

mile group in Fairbanks, was transferring down to the expanding operations staff in Anchorage to work on start-up plans. "I think we'll meet the schedule," said Anderson. "But there will be problems. We did the easy things the first year."

Kay Eliason was Moolin's second-in-command as manager of pipeline construction. The burly civil engineer had worked for ARCO and then in private consulting. He came to the project in Houston in 1972, first on the river and floodplain design and erosion control and then moving over to the field construction staff in Alaska when the haul road was built. "The second half of construction is an opportunity to start over . . . there aren't many projects where you can do that," said Eliason. "Balancing design and constructibility is always a grey area in a project of this kind, because you get conflicts between the engineers and the builders. So we'll have problems. Keystone Canyon and Thompson Pass, in Section One, are going to be difficult areas. We have some difficult floodplains to get through in Section Two, and some rougher terrain and longitudinal gradients in Section Three. And Atigun Pass, in Section Five, is going to take a lot of man-hours."

Moolin was also looking at the challenges ahead. "I'm concerned about places like Atigun," he admitted. "In some areas, we did eat our cake last year. But we did, in 1975, some of every kind of work that we'll have to do from now on. We do still have some physical exposures. The windows on our river crossings are extremely tight. Our biggest exposure is floodplain work, which is very difficult, with deep burial, and closely related to the weather we get . . . the amount of rainfall in the streams."

Soil conditions in Section Two between the Salcha River and Sourdough had just dictated the last major mode change to elevated line. "We thought we were exposed to 50 miles of re-mode in 1976, but it's going to be 24½ miles of additional above-ground construction," Moolin said. "We've ordered 26½ miles of materials for that, and we don't have any more lead time on ordering. We think the extra two miles of material is margin enough. If we need ten miles, we're in trouble. Oh, if we need a few hundred feet here and there, we'll find some solution. But if you see a mile of pipeline sitting up on wood blocks next year, or up on sticks like the Russians did it, you'll know somebody goofed."

Goofs were against the rules this year. Moolin's staff had used the winter layover to put together a detailed, inch-thick control manual of procedures to be followed through every operation. This was distributed to everyone with a role in building the mainline, including the thirty-one contractors and all of the unions. "Our contractors are as concerned about what's going on up here as we are, and we're seeing a lot more interest now on the part of their senior management and top executives," Moolin said. Midwinter meetings had also been held with the unions on work quality problems, and in some cases the unions were pushing quality standards now that were even stricter than Alyeska had specified in the first place. The control manual made news of its own. "Alyeska Attacks Waste," headlined the *Anchorage Times* one evening in January.

Quality control was one area the newspapers hadn't been paying much attention to. It was getting plenty within Alyeska itself. For all of the publicity about Arctic terrain and other Alaskan ground conditions during the project's environmental delays, some of the contractors had come to Alaska still insisting they could apply conventional pipeline construction wisdom to the work. From the start, Alyeska had given its quality control staff the authority to stop the work in the field when it wasn't being done correctly. And they were doing so. On one occasion over the Christmas holidays, a crew was held back from leave to redo some work, and its defiant foreman was finally fired off the project.

Robert J. Westerheid was Alyeska's manager of quality assurance and safety. "This project by dint of the federal government has a high quality control requirement. So when a guy says: 'It isn't according to spec but it's as good as we can do' . . . that isn't good enough on this project," said Westerheid. He called it the old-line pipeliner's "thirty-year" syndrome. "You know, he stands there cocking his hard hat up and spitting tobacco juice into the dirt, and he tells you "Shucks, son, I been building pipelines for nigh on to thirty years.' Well, you can't build this line the way you built lines in Texas and Oklahoma . . . just like you couldn't build an Apollo space capsule the way you built an F-86."

The comparison was deliberate. Before coming to Alyeska in 1975, Westerheid was with General Electric in Houston, heading a two-hundred man staff on GE's reliability and quality assurance contract for the Apollo program. Westerheid could draw a deadly serious analogy. "NASA had a problem . . . the Apollo fire. We're having problem . . . the welding radiography," he said. "In the Apollo fire, lives were lost . . . those three astronauts were my

1976 . . . Toward Completion

friends . . . and it cost millions of dollars in redesign.. On the pipeline, we have cost problems. Digging up pipe is expensive."

The welding situation had escalated into a full-blown controversy. The first problem was simply to define the problem. In taking back responsibility from its contractors, Alyeska had inherited a massive problem in radiographic record-keeping. The difficulty was in discovering how much of this was sloppiness, and how much was due to dishonesty on the part of subcontractors. After that, the problem lay in finding out whether only the records were faulty, or the welds themselves—and if so, how many welds, and how much repair work was needed.

It was another snarl born of the rush to move the project along in its first year. And Alyeska had been working to sort the mess out since mid-1975. Peter Kelley's suit against Ketchbaw Industries had been proof of contractor corruption in the welding X rays. In December a Ketchbaw project manager had been found dead in his Fairbanks apartment, and it was later ruled that he had taken cyanide. In February 358 radiographs of welds to be repaired were stolen from the construction camp at Delta in Section Two. They were never found, and it was never learned who took them.

Alyeska by now was doing its own review of all 30,800 welds made on the project during 1975. This was submitted to the Interior Department and to the Alaska state pipeline office in April. It was followed in May by Alyeska's technical analysis of the audit, and a report on repair work already under way and the various testing methods available to certify all past and future welding. All told, the project when finished would contain some 65,000 field welds from end to end of the mainline. Interior retained Arthur Anderson & Company, one of the largest U.S. accounting firms, to recheck Alyeska's audit. It also hired Southwest Research Institute in San Antonio and a Los Angeles metallurgist as additional experts. And Alyeska, through Harry Cotton at British Petroleum, arranged for more welding tests at the British Welding Institute in London and the nearby Cranfield Institute of Technology.

Alyeska's review produced a final list of 3,955 welds about which there was some question. The issue remained: Which ones were bad welds and which were just badly documented? Of the welds that showed flaws in X rays, the question was whether the flaws were serious enough to require a total repair. The list included 1,677 welds with flaws that the company felt were not great enough to pose a risk under the pipeline's operating conditions. There were another 517 lesser irregularities that Alyeska also felt were too slight to affect pipeline safety. All radiographs were numbered serially, and there were 360 missing or partially missing films in the series. That meant either an X ray had not yet been taken or that a weld had not yet been made where sections of the pipeline still had to be tied in to each other.

The problems here were many. Sorting out the records was a long process. Meanwhile it was mid-1976 and the work was pressing ahead to meet the year's completion goals. New X rays of almost 4,000 welds would take more time, especially since more than half of them were in already buried pipe, around which a deep "bell hole" would have to be excavated to allow testing and, if needed, rewelding. And some of these were under river crossings—the hardest of all to get at.

There was an argument over the standards that defined what was a good weld in the first place. The common standard was the American Petroleum Institute's Standard 1104. But it had been written thirty years ago, when small-diameter pipelines and less exotic materials were the norm. Also says BP's Harry Cotton: "The length of a fissure isn't important . . . it's the depth. The API code was silent on that one vital dimension . . . depth." The newer technique of testing by fracture mechanics, in which a weld sample is precisely notched and then bent to destruction by the constant application of a three-point force, had been highly used and accepted in the United Kingdom and Europe, and was considered to be more reliable than older testing methods still standard in the United States. It would prove, Alyeska claimed, that welds of the metallurgy and diameter on the pipeline were sound even if they did contain small and shallow flaws that the outdated API standard rejected.

By July, Congress had held the first of several hearings on the welding problems. President Ford sent a government team up to Alaska led by John W. Barnum, deputy secretary of the Department of Transportation, whose Office of Pipeline Safety was also monitoring the project. The officials reported that the welding records still needed a lot of straightening out, but could eventually be corrected and, in the future, improved. They were also convinced that the number of questionable welds was not any larger than already reported—and that extensive repairs were already under way.

In fact, the debates over testing standards and which welds might or might not have to be repaired were quickly becoming moot. Alyeska's deadlines couldn't wait for the arguments to be settled. By September, 3,175 of the questioned welds had already been reconfirmed as okay or done over again. Alyeska asked for waivers on 612 of the remainder, but the request only prompted more arguments and hearings. By the end of November, the problem had boiled down to 34 welds. John Barnum ordered 31 of them to be redone. Of these, 21 had been pronounced passable by the use, finally, of fracture mechanics tests. But since they were already dug up anyway, Barnum said to go ahead and fix them. Of the original list of 3,955 welds, waivers were granted on only three—buried seventeen feet under the Koyukuk River on the south side of the Brooks Range.

Alyeska estimated a cost of $55 million for the welding repair program in 1976. This was a part, but a minor part, of another 10 per cent hike in the project's overall cost. In July that figure was raised to $7.7 billion. Contributing to the increase, Alyeska said, were added material and freight costs, the repairs, and some contingency estimates made by projecting the past year's cost and design experiences.

In June Alyeska had gotten a new president. Ed Patton now stepped up to chairman and chief executive, and would continue in charge of the company. The incoming president, who would assume control after the line was in operation, was Dr. William J. Darch of British Petroleum, a Welsh-born chemical engineer who had just spent two years as general manager of BP's worldwide refining activities. Darch, by the age of fifty-three, had logged the sort of career that distinguishes an international oil manager: crude production and a liquefied natural gas plant in the Middle East, supply and marketing jobs in Australia, a refinery project in Africa, another one in Canada.

On a Saturday morning in late August, in slacks, an open-necked wool shirt, and sports jacket, Bill Darch was at his fifth-floor desk in Alyeska's Anchorage headquarters. At ease, he still smoked frequently, alternating from packs close at hand of Salems and Players. The desktop was a sea of notebooks and graph paper—"I've got a chart for everything"—as Darch talked about his new job: the start-up of the Alaska pipeline system.

"Theoretically, a pipeline from Prudhoe Bay to Valdez looks like a fairly straightforward thing," Darch was saying. "In fact, it isn't. This is not a fat old pipeline going across the desert." Between now and the middle of 1977, as a manager, Darch would be organizing and staffing Alyeska's next identity as an operating company. As an engineer, he would be working on the technical complexities of start-up itself: What happens when you first put a few hundred thousand barrels of very warm crude oil into a very cold pipeline and begin pushing it 800 miles over three mountain ranges toward its loading port.

Quite a lot could happen. Not only was the pipeline system brand-new, but its thermal extremes were greater than most pipelines ever encountered. During the start-up period, the system had to be brought into balance. "It will," said Darch, "take six months for the oil and the pipeline to reach a thermal equilibrium with each other." But Alyeska wanted to know, beforehand, what to expect during those first months. "This is all computer-simulated, and we've done 'test-tube' studies," Darch said. "But we're also building 8,000 feet of small-diameter pipeline . . . four inches, six inches, eight inches . . . on the North Slope for field tests that we can extrapolate from."

Meanwhile, there were less than four months of the working season left in 1976. "Right now," said Darch, "construction is still going on side by side with our start-up preparations." So it was. And right now, the pipeline builders were having just about every kind of trouble that their tortuous route through Alaska could throw at them.

The Bell Ranger II turbojet helicopter lifted gently off the pad at Galbraith Lake Camp, 150 miles above the Arctic Circle, 120 miles below Prudhoe Bay. Larry Schmidt, the pilot, headed low over a small pond by the camp, where two white whistling swans did not even raise their heads. The Bell straightened out over the pipeline route, holding altitude between 250 and 300 feet, following the land south as it began to rise toward Atigun Pass. On the other front seat to Schmidt's left, a husky passenger sat hunched forward for an unobstructed view down through the glass canopy, a clipboard on his lap, cigarette smoke drifting back past his face.

Kay Eliason was flying the line.

Back in January, in Fairbanks, Eliason had said: "Come back in the summer. I'll be going up and down the line by helicopter every week or ten days . . . you can see a lot of things you can't see from the ground." And you could. But this was not the sort of trip the pipeline construction manager had had in mind. It was a late afternoon in September now, and while the daylight lasted, Eliason wanted a firsthand look at the work from Galbraith over Atigun and down past the Dietrich and Coldfoot camps. This was Section Five, a combination of the two original northern sections, and there were serious construction problems over a good part of its more than two hundred miles from south of the continental divide up to the North Slope.

Larry Schmidt took the Bell Ranger up into the pass. A wolf foraging on the rocky slope below looked up briefly. The line was mostly elevated here, with an occasional arched overbend or a "dipsy doodle" animal crossing interrupting the level march of the VSMs. It curved to the right, ascending to its 4,790-foot maximum elevation in the pass. Eliason asked Schmidt to make a tight circle over Atigun, where work crews, their construction lights coming on in the dusk, were building what almost looked like a buried concrete bridge in a broad ditch through the saddle of the pass.

Out of Atigun, the helicopter followed the line closely again as it began to drop down the south side of the Brooks. Sections of buried line alternated with elevated portions, where the southerly sun glinted off the galvanized metal jacketing of the insulation. As the pairs of VSMs flicked by under the chopper, the thermal radiators atop their heat pipes looked like metal cactus flowers transplanted far from some Arizona desert. Periodically, every four or five miles on this gradual downslope of the route, one of the huge gate valves bulged up over the line.

Past Dietrich and almost to Coldfoot, Eliason signaled his pilot to turn back toward Galbraith. "Okay, I've seen all I need to see," he said, leaning back in his seat for the first time in an hour. He had seen almost one hundred miles of the route. The top page on his clipboard pad was covered with notes.

Kay Eliason had moved up to Galbraith from Fairbanks in July, bringing with him Robert M. Smythe, an organization specialist from the Wainwright staff. The two had gone north to reorganize the efforts in Section Five. "Nothin' was working right up here . . . there was just a lack of a sense of direction in the section," said Eliason.

Frank Moolin was also at Galbraith Lake now. Early this year, *Engineering News-Record* magazine had named him its 1976 "Construction's Man of the Year." Moolin was earning the honor all over again. Alyeska had moved the pipeline department headquarters into the heart of its worst trouble spot. "We're up against a helluva lot of problems, and we've got an extremely difficult time schedule now," Moolin conceded. "I guess, if I had to say it in hindsight, the biggest mistake we made on the project was having one executive contractor for these two sections. And then we assigned them the fuel line from Prudhoe Bay on top of that."

The problems were a combination of weather, terrain, contractor performance, and the welding repairs. Moolin added a fifth: "Getting everybody to stop pointing fingers and make this temporary organization work." Too many people were blaming the welds. Section Five, it turned out, had almost half of the questioned welds, a large part of them below ground. Even by September, said Moolin: "We've got 1,003 . . . 1-0-0-3! . . . locations to excavate, X-ray, verify or repair, and backfill. It's a nightmare of coordination."

The welds were only the start. At a buried crossing under the Sagavanirktok River near Happy Valley, two camps north of Galbraith, 1,700 feet of concrete-coated pipe had floated up out of its 20-foot-deep trench in June. The trench had not been backfilled because the fill material had frozen solid, and the heavy spring flows over the floodplain of the Sag had battered the concrete jacketing and flattened the thick-walled pipe into an oval shape. The pipe was replaced and the crossing was reengineered into another channel. Since this was not regulation summer work, said Eliason, "the Fish and Game people gave us twenty-four hours to build it . . . we did it in four hours." Near Toolik, the next camp over, some construction in a cramped area was permitted only from a snow pad—and there was no snow yet. But later in the year, when there would be snow, the temperatures might be too low for welding.

Then there was Atigun Pass. This summer, glacial soils had been found in the original burial route. The area was also exposed to frequent avalanches, so elevating the line here could never be considered. Another 180 bore holes were drilled, and the angle of the route was shifted several times. The solution finally was to design a structural, insulated ditch—a 6,000-foot-long, 8-foot-square concrete box with the pipe buried inside it wrapped in a 21-inch thickness of high strength Styrofoam. The task now was to build this before the end of the season. "I think we're okay on that . . . unless we get three feet of snow or a gullywasher this month," said Moolin.

The real pressure came from the cumulative delays of all of these problems. This would hold up the year's main goal: to get the mainline installed and hydrostatically tested before winter closed in. Hydro-testing was standard practice in pipelining—the ultimate test of a line's structural integrity. In Alaska the mainline was being tested to 125 percent of its maximum operating pressures. Short sections of welded pipe were sealed off and water was pumped in at pressures of up to 1,800 pounds per square inch.

The last step before hydro-testing was the accurate alignment of all elevated pipe—making sure that every bend, and the degree of each bend, was correctly positioned so that the pipe crossed each support exactly where the design said it should. Alignment was now being checked over every portion of the more than 400 miles of above-ground line. At Galbraith Lake, grim-faced in September, Moolin and Eliason looked as though they wanted to go out and yank the entire pipeline into alignment by themselves. "Both of these guys are really under pressure," said Bob Smythe. "You'd think either of them could just fall over at any time."

Section Five had more than its share of problems. But it didn't have all of the problems. The hydro-testing was proving out one after another section of the line without a hitch. But in July, during pressurization of one 480-foot segment, seven feet of the pipe wall had ruptured and a short length of the pipe had expanded in diameter. Usually a pipeline's girth welds are its strongest points, and the longitudinal welds are the next strongest—the wall itself being relatively the weakest. Human error and a faulty gauge, it was found, had put several times the maximum pressure into the section. All of the pipe was replaced, rewelded, and buried again.

Down in the Chugach, meanwhile, two more chunks of Alaskan terrain only a few miles apart were bringing out all of the ingenuity that the engineers and builders could come up with.

The work at Keystone Canyon had stopped in 1975 with most of the right-of-way done. It started up again in May. The last of the three hundred inches of snow that had fallen over the Valdez area during the winter was cleared off after it had been sprayed with carbon black to speed its melting. Crews now began ditching the high canyon edge for the pipe burial. That wasn't easy. More than half a million cubic yards of rock had to be blasted out to make the ditch—120,000 yards of it at one spot with a 60 percent grade. For the gravel pipe bedding, a rock-crushing plant was built atop the canyon to produce some 40,000 yards of bedding material that otherwise would have to be trucked up the switchback road to the top.

Getting the 80-foot lengths of pipe up to the ditch was the rough part. The sections were too heavy for helicopters, and the regular pipe-carrying trucks couldn't manage the road. Alyeska and MK-River finally rebuilt a bulldozer with heavy crossbars and cradles to carry a pipe section along each side. Each load took half an hour to navigate the switchbacks, with a second bulldozer cabled to the first and going ahead of it as a helper. A full week was spent getting all of the pipe up there. With that done, the pipelaying itself was almost an anticlimax.

Keystone Canyon was finished and hydro-tested by November. The section was just over four miles long. But its 23,000 feet of pipeline had to rise more than 800 feet to get into the canyon, and drop another 800 feet to get out, and in between the line went through more contortions than could be found on any other short span along the route. For all of this, tourists who stopped on the highway below to watch the crystalline steam of Bridal Veil Falls, the area's prime attraction, never knew what was going on above them. The builders had been collecting the silted water in catch-basins to let it settle before it went over the waterfall.

Thompson Pass, north of Keystone, was another exercise in what seemed almost like vertical pipelining. This divide, 2,500 feet up in the Chugach, ran for several miles and confronted the crews with slopes up to 45° over which the pipeline had to be ditched, laid, and backfilled. Heavy construction machinery was held with anchored cables to keep it from plummeting off the slopes. The pipe itself was winched up each side of the pass, length by length, with a cable tramway system. Welders harnessed themselves to the pipe itself, as they worked on it, to keep their footing.

Thompson, too, was only a few miles long. But it took the

entire 1976 construction season to complete. Ditching was not finished until mid-October, as the area's typically early rain and snow and high winds slowed down the work crews. A week or so later, pipe installation over Thompson Pass was completed as a 3,700-foot section on the steep south slope was tied into the line running south to Valdez. In worsening weather, backfilling of the pipe ditch continued into December, and the hydro-testing at Thompson Pass was finished at mid-month.

Work at the pump stations had gone on at a good clip all year. It had started with a total redesign of the site foundations at Pump Station 6, just south of the Yukon. Borings there had shown the soils to be stable, but the excavation and grading work revealed ice-rich permafrost lower down. The site was redone to use the buried refrigeration coils already in the foundations of four other stations. "So we had a wintertime cement pouring operation up there," said Rod Higgins. "You wouldn't do that again if you had your choice."

Month by month, the installation of turbines, pumps, valves, and piping continued at the first five stations needed for starting up the line, and at Station 5, the surge-relief station on the south side of the Brooks that was being finished except for mainline power. By September, these stations were being hydro-tested, and the first operating specialists were arriving to begin checking out and debugging the station subsystems. Each station was to be tested internally as a "closed loop," sealed off from the mainline, by pumping Arctic diesel oil around and around its systems so that any problems would be located before the station was tied into the mainline in 1977. "It's very similar to your new automobile," Higgins said. "You drive it around for ten days so that any components that are going to fail are given time to fail." Three more stations needed after start-up were brought along on the same sort of schedule, to be finished by year-end. The last three stations, not needed until the line was boosted to full capacity in a few years, were being finished as pass-through stations without buildings or turbines and pumps.

The Valdez terminal, by September, was two-thirds complete, with 4,200 workers pushing ahead on every part of the complicated facility. "Nothing that we're building here is all that new," said Vic Filimon, Alyeska's manager of terminal construction. "It's the size and the logistics . . . the bigness of the place and the strictness of the specifications. Everything here is being built to withstand 90-pound snow loads, 90-mile-per-hour winds, and an earthquake of 8.5 on the Richter scale."

Filimon was a 30-year SOHIO veteran who had supervised refinery, chemical plant, and terminal projects in Iran and around the United States. "The thing that's given us the most surprises here has been the earth-moving we've had to do . . . because of what I call the 'consistency of the inconsistency' of the rock," Filimon said. "We're moving 14-15 million cubic yards of rock and dirt, which is an awful lot. We expected the ring walls that the tanks rest on to go down eight feet . . . some had to go forty-five feet to find bedrock."

Donald R. Roberts, another SOHIO construction manager, was directly supervising the earth work and other civil engineering throughout the terminal. A big part of this in 1976 was installing a system of retaining walls against the terraces that stepped up from the bay. "We thought the cuts and fills would be easy," Don Roberts said. "Instead we found there was more overburden and glacial till than we expected, and the rock was of poor quality. The slope angles would have had to run right down to the bay, so these walls conserve a lot of area." The walls, a technology developed by Reinforced Earth Company of Seattle, were made of 25-square-foot hexagonal precast concrete panels, dovetailed together and anchored by galvanized metal strips that extended from each panel into the filled ground behind the wall to a distance roughly as far back as the wall was high.

A total of fifty-eight structures, including tankage, were to be finished within the Valdez site by the end of this year. The project manager on these was John R. A. Sutton, a BP pipeline engineer who had come to Alyeska from the North Sea by way of the terminal design staff at Fluor in Anaheim. Springing along the catwalks of the power plant and vapor recovery buildings, Jack Sutton pointed out the 300-foot boiler stack that had been pivoted into place in one piece this summer. "The prime purpose of this power plant is not to generate power," Sutton explained, somewhat tongue in cheek. "It's really to generate the inert flue gasses to go on top of the oil in the tanks."

Sutton's biggest task now was to complete all outside work and hydro-testing of the terminal's piping and tanks before winter.

The water for these pressure tests came from one of the several streams that ran down out of the mountains and through or around the terminal. It was cleaned after the tests and pumped back into another stream. "The testing pressures aren't particularly high," Sutton said. "The maximum design pressure with oil in the tanks is 72 percent of what the tank wall thicknesses can actually take, and we test them to 95 percent."

At the terminal's shoreline, the two tanker berths needed for start-up were more than half finished, and two more would be finished in mid-1977. Anchoring the berth pilings in the rock under Valdez again was a challenge. "We needed to go on additional sixty-two feet to find bedrock for some of these piles," said Marlin H. Leinbach from ARCO Pipeline, a project engineer on the offshore facilities. Work on the berths was being pushed along before water temperatures dropped and winter winds rose to hamper the work. "The winds have favored us so far," Leinbach said. "They can be something else up here."

Through the fall of 1976, despite weather and everything else that could work against the project, most of the year's problems were overcome somehow. During the first week of November, the first ship docked at Valdez. The *Toshin Maru* from Japan tied up at the completed Berth 4 to unload more pilings and other steelwork for the unfinished berths. The first two computers had been installed in the operations control center, and in December the first operating command was sent out along the 800-mile backbone communications system to test a mainline valve at Pump Station 2 on the North Slope. Along the route, the first nine pump stations were essentially complete, with their tankage hydro-tested, and testing of their mainline systems and functions still going on.

The critical problems on the mainline itself, especially in Section Five, one by one had been just about whipped. At Atigun Pass, with a break in the weather, backfilling of the buried concrete box structure was finished in late November. The opposite kind of weather break—a good northern snowfall in October—provided the urgently needed snow pad for construction near Toolik Camp. All 78,000 VSMs had been in place since midsummer—in November, the last of the 62,000 pairs of thermal heat pipes were installed.

The final piece of all of the 800 miles of mainline pipe was installed on December 6 near Thompson Pass. All that remained to create an uninterrupted pipeline were the tie-in welds after the last sections were hydro-tested. Weather, the weld repairs, and the troubles in Section Five had defeated Alyeska's goal of having the line totally finished and pressure-tested this fall. At year-end, 160 miles of hydrostatic testing were left to be done when Alaska warmed up again in 1977. About 60 miles of above-ground pipe still had to be insulated. There were thirty-three weld repairs yet to be made. And rocky ground had slowed the ditching for the 148-mile gas fuel line to the first four pump stations, leaving 78 miles yet to be installed in 1977.

All told, at the end of 1976, the Alaska pipeline system was 92 percent complete. The mainline, at 97.5 percent, was all but done. Sections One through Four were virtually wrapped up except for final touches and clean-up, while Section Five had the only real construction work left. The pump stations were 92 percent complete. The terminal stood at 83 percent now, and almost its full work force would keep at it all winter. Total project employment had peaked above 21,000 workers again in 1976, and now had dropped down to about 6,000 during the winter layover.

Frank Moolin and Kay Eliason were keeping their base at Galbraith Lake until Section Five was all done. Moolin was too experienced in construction to let himself get overconfident— especially on this project. But a lot had been accomplished since his department was redeployed up north last summer, and he was smiling again. "Well, we set some goals and, God help us, we met them," Moolin could say. "So we did it . . . we'll finish on time next year. There were a lot of doomsayers around, but I guess we just didn't know it couldn't be done."

15.
The Environment...
Putting It All Back

In Atigun Pass, on the sides of the earth cuts above the Alaska pipeline's special concrete-box burial, new autumn grass is poking through coarse excelsior mats laid over the recently graded slopes.

On the flatlands near Dietrich Camp, and in hundreds of other locales where the almost finished line crosses reasonably level terrain, the ground beside the work pad is greening up again with thick, wind-ruffled swaths of winter rye.

Near the Sagavanirktok River, where a difficult river crossing has just been rebuilt, an Alyeska biologist is checking stream velocities and fish passages during the annual migrations of Arctic char and grayling.

At the Gulkana River, salmon are running beneath a new 400-foot tied-arch bridge where Alyeska first planned to bury the pipeline so that it would not intrude on the area's scenic vistas from the heavily traveled Richardson Highway nearby.

From Prudhoe Bay to Valdez, there's not a mile of the pipeline route that hasn't been environmentally engineered in some way. Over the years of the project, Alaska's ecology and the oil line have become so intermixed that it is hard now to tell which is more a part of the other. From 1968 on—long before the NEPA statute or the lawsuits that sprang from it—the project's planning had an environmental quotient as unique to pipelining as Alaska's natural landscape is to the rest of the United States.

Long-distance pipelines in the Lower 48 have faced environmental challenges every time they crossed places like the lakelands of upper Michigan or the sheep ranches of west Texas. "Now, this is something that every industry faces," says Frank R. Fisher, Alyeska's manager of environmental protection. But Alaska, as in all things, was a different place to begin with. And so the response to its ecology was different from the start. Fisher, a chemical engineer at ARCO's production research laboratories in Plano, Texas, was given the job of coordinating the company's Arctic research after the North Slope discoveries, and came to Alyeska in 1973.

Fisher's department, from the beginning, has been assigned to the company's operations division, since it will continue to be a part of Alyeska as an operating company. But the department has really spanned the project from construction management through to the plans for start-up, expanding its functions as it went along. It began with two sections. Fish and wildlife administration was under Ben L. Hilliker, who had been with the state of Alaska until 1972 as a deputy commissioner in the Department of Fish and Game. Air and water quality were under Dale E. Brandon, an oceanographer from Exxon Production Research in Houston. In 1976 the department added restoration, revegetation, and erosion control to its responsibilities, bringing Al Condo over from Hal Peyton's staff as supervisor of restoration engineering. A contingency planning staff was also set up to handle oil spills, earthquake damage, and the like—this was under Frank Therrell, who also continued as manager of project permissions.

In all of these areas, North Slope oil and the pipeline have inspired—and funded—a lot of Arctic research that might not otherwise have got done for a long time. "This project has generated and gathered together a lot of good environmental information," says Ben Hilliker. "It's stimulated a lot of work that was going to be done at some point . . . but it's speeded it up maybe by twenty or twenty-five years."

The environmental programs were planned in three stages. Soil stabilization, erosion control, and some year-to-year restoration and revegetation work were the most important tasks during the construction years. During and right after the start-up period, the main work has been restoring and replanting the entire pipeline route. From then on, maintenance of the pipeline system will include a perpetual watch on the environment through which the line passes. A part of all this has been a visual impact program—the "cosmetics" of the pipeline. Pump stations, as much as possible, were designed and painted to blend into their surrounding countrysides. Tree plantings and other techniques of landscape architecture are being used to make any portions of the line in public view simply look better—sometimes better than many areas had looked to begin with.

The research that went into these programs began during the preconstruction period. It was just as intensive as the research going into the pipeline design itself. Over its 800-mile route, pipeline construction would affect 30,000 acres of Alaska. This acreage was classified into seven physiographic provinces—vastly differing from Prudhoe Bay's Arctic coastal plain to the heavily forested slopes of the Chugach, with everything in between from the tundra and glacial valleys of the Brooks Range to the floodplains of the Copper River basin and the central interior. There was also great variety in the landforms within each province.

While the pipeline engineers were boring into the earth of Alaska for structural data, Alyeska's agronomists were taking surface soil samples—15,000 of them—along the route. These were organized into 826 composites to be chemically analyzed. There were a number of questions. What grew in each kind of soil? What else would grow there? What kinds of fertilizer worked best in each soil? How much moisture was each type of soil used to? Precipitation varied between one hundred inches per year at the pipeline's southern end and only four to eight inches a year at the North Slope. So what would be the erosion characteristics of each type of soil in each area, and what methods would work best in each area to control erosion?

Reseeding the pipeline route was not as easy as it sounded. No one had ever bothered to reproduce commercial supplies of seed for Alaska's unique natural vegetation. "We probably tested more than 450 winter-variety grasses from all over the world before we were through," Al Condo says. In the end, at the twenty-six test plots up and down the route, only a handful would "take" in Alaska. They had names like boreal red fescue, arctared fescue, manchar brome, meadow foxtail, and climax timothy. Four different mixes were put together that grew successfully in different regions of the route. Each included a batch of winter rye to stabilize the slower-growing varieties that would seed the longer-lasting growth. The same sort of experimentation was carried out with five different fertilizer formulae to be used in areas where they were most effective.

Then the agronomists had to classify the route again by growing seasons. In the south, seed could be planted from mid-May until August. From Toolik north to Prudhoe Bay, the sowing season lasted only from early June through mid-July. Seeding in late summer would cause winter kill with the first frosts. But dormant seeding could be used after each year's first hard frost, with the seed staying in place all winter and sprouting after the first spring thaws the next year.

The purpose of these programs was to firm up the ground around the construction even as it continued from year to year. "Stabilization of the pipeline comes first . . . we don't want it to wash out," says Al Condo. He points out that much of the seeding and fertilization program, in fact, was temporary—meant to go on only for the first few years until the natural revegetation reinvaded each area. "A Band-Aid, if you will," Condo says. "We know that Mother Nature is the best erosion-control engineer of them all."

Erosion control during the construction years went well past revegetation. The pipeline work pad itself was engineered as carefully as a superhighway through the tundra would have been. Gravel thicknesses, grades, and embankment angles were individually worked out for soil and weather conditions in each area. North of the Brooks Range, seventy miles of the work pad were synthetically insulated with 94 million board feet of polystyrene foam, one and a half to three and a half inches thick, laid down over a six-inch gravel bedding and covered over with another sixteen inches of compacted gravel overlay. One result was that, since the work pad along this mileage was reduced from thicknesses of as much as seven feet to only twenty-four inches, millions of cubic yards less gravel had to be taken from material sites.

Erosion control was also a factor in the design of each crossing over the dozens of rivers and hundreds of streams along the pipeline route. Erosion and siltation are natural phenomena—they had been occurring and curing themselves for centuries in Alaska before industry came into the wilderness. But the building of the pipeline was a temporary intrusion that would aggravate nature's processes, and this had to be held to a minimum. This meant a specific design of each culvert and low-water crossing, armored with a rocky bed, at each place where a stream crossed the pipeline work pad. These crossings were also designed to allow the normal fish passage in each area. Stream beds themselves were reengineered where the construction changed their channel structures or gradients. Plunge basins and 'graduated "let-down" structures were inserted to avoid too much velocity building up in a stream as a result of a culvert or crossing. Levees were also built to cut down stream velocities and to divert stream flows away from a cut slope.

Alaska's widespread and unpredictable groundwater conditions were another erosion-control target. Impermeable plugs were used in the pipe ditch itself to keep groundwater from washing away the pipeline bedding by diverting it outward from the ditch. Damage to the tundra overlaying permafrost would also result in erosion and creation of groundwater flows. So the erosion-control programs included restrictions on use of various kinds of wheeled and tracked vehicles, in favor of roller-equipped "rolligon" vehicles, over the tundra. There were also procedures for repairing damaged tundra: reinsulating the permafrost, but not overinsulating it so that it could not refreeze by its normal cycles. And there were methods for dealing with the insidious *aufeis*, keeping it from climbing up over the pipeline route by damming or fencing it in place until it melted again into a natural runoff.

Alyeska's environmental specialists were posted at each of the pipeline camps during construction. At Galbraith Lake during the hectic summer of 1976, biologist Kenneth Durley ranged over the line all the way up to the North Slope, keeping a particular eye on the river work in progress. At the troubled Sag River crossing, he said: "Moving the channel wouldn't bother the char and grayling migrations . . . in the fall, they are coming downstream. It would have been a problem in the spring, when they're coming upstream to spawn."

Durley, who had been with Alaska's Fish and Game Department before coming on the project, also kept an eye on the four-legged wildlife in northern Alaska. Precautions had to be taken in spring with the Dall sheep, who came lower on the mountain slopes during the lambing season. Wolves were more leery of the construction activity, but would appear cautiously on the camp fringes and along the haul road, where workers tried to feed them. Bears, including sows and their cubs, simply walked through the camps at will, looking for food. Alyeska on occasion had airlifted them a hundred miles away, only to have them find their way back to the same camps. "We're not giving the bears any problems . . . they're giving *us* problems," said Ken Durley.

Mark Welker, a geologist who had worked at the North Slope in 1969, was the restoration specialist at Galbraith Lake that summer. Most of the current revegetation work, he said, was to head off any hydrolic washing away before the soils were restabilized. The excelsior webbing at Atigun Pass was to slow the melting in an ice cut—it could be peeled up periodically and laid down again after new seed and fertilizer were added. "At our Prudhoe Bay test sites," Welker said, "our revegetation is just about gone and natural reinvasion has taken over again."

The North Slope had been a mess in the prediscovery years, Welker said, because the military construction crews and later the first exploration contractors hadn't bothered to make the effort to cart away their empty oil drums and other debris. "There is nothing up here that can't be repaired with time and money, if someone will spend it," Welker said. "You hear a lot about the north of Alaska being environmentally sensitive . . . well, any place is environmentally sensitive. But a lot of environmentalists have seen by now that the environment here isn't as delicate as they first said. We've proven that things like permafrost can be stabilized."

Geoff Larminie was now back in London as general manager of British Petroleum's worldwide Environmental Control Center. When he was seventeen, Larminie related, he had read Spencer Chapman's *The Jungle Is Neutral*. The idea had stayed with him as he got his degrees in geology and zoology and began packing his curiosity off to the remote edges of the earth on BP assignments. "I've always felt the same way about the Arctic . . . that it is neither for you or against you," Larminie said. "You must never underestimate it . . . but by understanding it, you can use it for your purposes, and you arrive at a sort of rapport with it."

The environment, Larminie felt, was a total package—natural, political, social, economic. And it was often this very package of attributes that attracted people to isolated, difficult, seemingly hostile places. "There's always a mystique attached to the Arctic . . . the remoteness, and the relatively unexplored quality of the area, and people's ideas of the cold and privations that you must endure in an environment like this," said Larminie. "You feel that in some way it ought to produce finer feelings in people."

In fact, he felt, it does. Development, paradoxically, is drawn to distant wildernesses like the Arctic. It is a part of man's constant quest for challenges. But often those who come there are themselves changed more than any change they bring with them. "The hardheaded will say they do it for money," Larminie said. "But when you get to know people, and talk in the evening over coffee after a shift is over . . . in some curious, inarticulate way they're trying to tell you that there's something else that they can't understand, a fascination that they can't put into words . . . but it drives them up north too."

British Petroleum, born in the Iranian desert at the start of the century, today is seeking for oil—and often finding it—all across the northern top of the world, from Cape Lisburne and the North Slope in Alaska to Ellesmere Island and the Labrador Sea in Canada, and across to Greenland and the North Sea. "We've all grown up in that environment . . . we're fascinated by it . . . we love it . . . we understand it to a degree . . . and we want to see it looked after," said Geoff Larminie. "We know that we're going to develop it in an orderly fashion, and not make a mess of the oil fields . . . this is just not compatible with good engineering."

Robert O. Anderson, board chairman and chief executive of Atlantic Richfield, is cochairman of the International Institute for Environmental Affairs and a member of the National Associates Board of the Smithsonian Institution. These are not token tasks. Before 1968, when the ARCO/Humble discovery made a headline name of an almost nameless place, the North Slope had indeed been badly littered. And on a coastal plain that is frozen solid for much of the year, debris does not decay—it stays there forever. North Slope legend has it that the clean-up campaign began the day Anderson first flew there. British Petroleum, too, was flying out planeloads of other people's oil drums to Fairbanks from Prudhoe Bay.

The Slope today, for all that it has become an industrial park of vigorous oil operations, is a neat industrial park. A visitor from the litter-addicted Lower 48 will be startled to see the crew of a pickup truck stop abruptly on the oil field's spine road, in the half-light of January with the temperature dropping toward minus 50°, and scramble frantically over the snowy tundra to retrieve empty packing cartons that had blown out of the truck bed. Workers on the North Slope have been ordered to pack up and leave for not being as careful as this.

Angus Gavin is a Scotland-born naturalist who took his degrees in zoology and engineering and moved to Canada. After a career of more than forty-five years, much of it in the Canadian Arctic studying the wildlife and working with Canada's Eskimos there, Gavin in 1969 was asked by Atlantic Richfield to come out of retirement for what turned out to be a six-year study, from 1969 through 1974, of the North Slope's ecology and the impact of oil operations on it. Gavin's work added much to the pool of environmental data being gathered from end to end of the project. His overall conclusion at the North Slope is that nature often shows more resilience and adaptability to man than man does, or sometimes is willing to do, to nature. "After five years of oil operation activity on the North Slope," Gavin wrote in 1972, "it is quite evident that these operations have had little or no effect on the normal activities of the wildlife that frequent this part of the Slope."

The growing traffic at Prudhoe Bay—from trucks, helicopters, and arriving and departing freight and commercial aircraft—seemed to be making little difference to Alaska's own creatures there. From the start, no one came back from the Slope at any time of year without at least one wildlife story. Caribou were grazing placidly beside the humming airstrips. As the oil companies raised their base camp buildings on pilings above the tender tundra, caribou used the air spaces beneath for shelter from Prudhoe Bay's long hours of summer sun and notorious swarms of flies. Alaskan grizzlies and brown bear raided camp garbage pits, quickly bringing about stricter housekeeping rules. Starting in 1969, Arctic white fox and red fox were darting among the mounting stacks of pipe arriving from Japan.

This polar desert by the Arctic Ocean has been the immemorial summer destination of Alaska's great migrating herds of caribou. Man may have come to the Slope now. But so, still, do the caribou. Tens of thousands were counted in 1969 and 1970. Angus

Gavin's team of naturalists was worried when the 1971 and 1972 migrations seemed to drop to a tenth that many. But the next year the herds again appeared in numbers approaching 20,000. The answer was that heavy winds and snow conditions in the Brooks Range in the winters of '71 and '72 had decided how far north the caribou could range the next summers—not the presence of the oilmen.

Waterfowl—geese, swan, loon, tern, and many others—also kept up their annual migrations to and through the North Slope. "Until 1971, there were no snow goose colonies in this area," Angus Gavin related toward the end of his studies. "Yet, despite all the air traffic in and around the field, a colony of some fifty pairs started nesting on Howe Island at the mouth of the Sagavanirktok River. If helicopter noise and overflights were a factor, I am sure this colony could have found a far quieter place, either east or west of this busy field." Yet, said Gavin: "This colony has come back each year and, in 1973, had increased to some sixty-odd pairs."

16.
Prudhoe Bay... Producing Oil and Looking for More

On the enclosed drill floor of Rowan Companies' Rig 26, tool-pusher Joe E. Hamilton and the rest of the crew were enjoying a climate no worse than south Texas on a bad night in January. It was only when the drill string was going in or coming back out of the hole and someone had to go topside into the exposed V-notch to handle pipe that reality set in. It was January, all right. But it was Prudhoe Bay, not south Texas, and the weather was a windy 50° below zero. And despite the pervading darkness all about the busy rig, it was just past four o'clock in the afternoon.

The rig was making hole now, and the man who scuttled up top as each pipe length was threaded into the one below it in the kelly was well wrapped for his work. Down on the floor, the rig's clanking diesel engine sent off vapors of condensing steam and the smells of hot metal and grease. The mélange of noise and odor made a phantasma of warmth in which the quick-moving crew could work comfortably in coveralls.

This was Atlantic Richfield's drill site No. 4 at Prudhoe Bay, and this was ARCO well No. 7 on the site. Through the gloom on the whitened tundra below, abaft of the working rig like a picket line of icy polar soldiers, stood the upright "Christmas trees" of the site's first six completed wells. These were the wellhead valves and outlets, shut in for now until Pump Station 1 and the pipeline eight miles to the west were ready to accept the first North Slope oil in 1977.

Rowan Companies had four rigs working under long-term contract to ARCO up here this winter. By 1977 there would be six rigs punching away at the known oil pools deep beneath this eastern half of the sprawling 300-square-mile Prudhoe Bay oil field. This was field development by ARCO operating for itself and Exxon USA. On the western side, where the other half of the field was being developed by BP Alaska for British Petroleum and SOHIO, Brinkerhoff Drilling Company and Nabors Drilling Ltd. were also pushing down wells. Parker Drilling Company had a wildcat rig for Mobil Oil, whose partners in the field were Phillips Petroleum and Standard Oil of Indiana. Yet another producer group on the Slope was a six-company combine of Amerada Hess, Getty Oil, Marathon Oil, Placid Oil, Hunt Oil, and Louisiana Land and Exploration Company.

The Prudhoe Bay oil field was to be unitized under a detailed agreement already being worked out between all of the companies operating there. This was so that the oil could be extracted with maximum efficiency and allotted to each firm on the basis of its proven leases. Prudhoe was never to be a tight-packed sea of wellheads and drilling derricks jammed out in all directions like the old-time oil fields of east Texas and California. Wells were being drilled in clusters from insulated gravel pads as large as 566 yards long and 78 yards wide, spaced out a mile or two from each other through each operator's zone. Six or eight wells were being drilled from each pad, with each well kicked out directionally at a bending angle in the drill string to tap a 640-acre expanse of the oil pool below. The angling meant that some wells were reaching 11,000 to 13,000 feet into the pocketed subterranean sandstones that held the oil.

By mid-1977, when the pipeline tapped into the waiting production, a total of some 150 wells were to be completed throughout the two halves of the field. Already, early in 1977, BP and SOHIO had completed 64 of their wells, and ARCO and Exxon had 59 of their development wells down. The target of 150 wells, with an average expected output of 10,000 barrels of oil per day from each well, was pegged to meet comfortably the initial pipeline throughput demands.

Each well was taking about thirty days to drill. After it was completed, its rig was skidded a short distance along the pad to the next location, where it could be drilling again within forty-eight hours. When a pad had its full quota of completed wells, the rig was trucked over the Prudhoe Bay spine road system to its next pad. Folding "jackknife" rigs could be operating again in ten days. Newer Arctic rigs, designed to come apart in sections with much of their gear left in place, could be into a new well in two or three days.

The wells themselves were engineered for the Arctic. Where the first two thousand or so feet of a well penetrated the North Slope's extraordinarily deep permafrost, the conventional single casing around the drill string was instead a double casing with a layer of thermal insulation—a General Electric development from the Apollo program—between the two casing strings. Heavier casing was also used below the permafrost in the event that the frosty subsoil slumped or migrated lower in the hole. To define its particular ground conditions all the way down, each well was electronically mapped and analyzed as it descended.

Unlike many older oil fileds where the familiar walking-beam pumping mechanisms must be used to bring up the oil throughout the working life of each well, the reservoir pressures at Prudhoe Bay are strong enough to produce oil almost like turning on a faucet. Each well was being completed with dual safety valves, one as part of its Christmas tree above the pad surface, the other set in the casing below the permafrost level. As soon as the pipeline was ready to take oil, drilling rigs were wheeled back to each shut-in well and special oil field tooling was sent down to perforate the production string at the level of each pool so that the oil could begin flowing.

Gathering the oil in from the production pads around the field toward Pump Station 1 calls for yet another pipeline system within the oil field itself. Eventually, this system will total hundreds of miles of pipe ranging up to forty-two inches in diameter as it approaches the pipeline's origin station. All of the piping in the gathering system is raised above the field and insulated with jacketed three-inch thicknesses of foam. On the ARCO/Exxon side of the field, three flow stations are consolidating and processing the oil before it goes on to P.S. 1. Oil from each cluster of wells on a pad goes through one common line from the pad to a flow station.

Each flow station has a capacity of about 300,000 barrels a day, handling the production from about twenty-five wells. the BP/SOHIO field system is almost identical, except that BP Alaska prefers to run a single line from each individual well, and its processing points are called gathering centers rather than flow stations.

Whatever they're called, the facilities have important tasks to do before the oil can be sent south. As usual in a producing oil field, some of Prudhoe Bay's immense reserves of natural gas comes out of the wells with the oil. This must be separated out before the oil can go into the pipeline, and so must small amounts of water that are also in the oil. The water is treated to take out impurities and then is reinjected into disposal wells at each central facility. The natural gas goes into an eleven-mile large-diameter field pipeline headed for Prudhoe Bay's single gas compressor plant on the ARCO/Exxon side of the field. Here, most of it is compressed and reinjected back down into the massive cap of gas that overlays the oil pools. And there it will stay with the rest of the North Slope's 26 trillion cubic feet of proven natural gas deposits until government decisions are made on the route of another giant cross-country pipeline system to bring it to market. When this is decided, the Prudhoe Bay compressor plant will be used to start it on its way.

Some of the North Slope's oil and gas are used to operate the field itself. Prudhoe Bay's single 134-megawatt power plant, which rises to the northwest of Pump Station 1, is fueled by the rest of the natural gas from the wells that has not been reinjected. The plant supplies electricity to all of the production facilities at the Slope and to the temporary construction camps there. Prudhoe Bay oil, for its part, is the feedstock for the first permanent facility built at the head of the pipeline route in 1969—a 13,500-barrell-per-day topping plant. This miniature refinery, close to the ARCO/Exxon base camp, produces the 3,500 daily barrels of fuels used at the Slope—the diesel oil and naphtha gasoline that powers all vehicles there. The residue of the refinery's daily throughput goes back into the reservoir.

Prudhoe Bay's other largest facilities are the BP/SOHIO and ARCO/Exxon base camps and operations centers. "Camps" is too modest a term, for these are permanent, not-so-small-scale office-hotel complexes, built to provide comfort and efficient working space to the operating staffs that will inhabit this far-off Arctic oil production center after much of the excitement has died down and both the field and the pipeline are into long years of normal operations.

The three-story BP/SOHIO center, which has been nicknamed everything from the "BP Hilton" to the "Ice Palace," was expanded in 1976 from 90,000 to 137,400 square feet, and now gives living and working room to 264 people. Since it opened in 1974, the incongruity of its brightly colored interior architecture has attracted magazine coverage worthy of a suburban estate. The BP center has glassed-in areas where something approximating "outdoor" sports and recreation can go on, and a glassed-in arboretum planted with birch trees and tundra flowers. There is also a 40-foot heated swimming pool and sauna bath that is utilitarian besides being recreational—the pool doubles as a reservoir in case of fire.

ARCO/Exxon's two-story center, eight miles to the east as the snow goose flies but a bit longer by the spine road, sets more examples in how to live if you have to live in the Arctic. First built in 1970, it now holds up to 440 people and its recreational perquisites include a small theater and a paneled indoor basketball court. On a more temporary basis, ARCO has also built a 1,750-person construction camp for the workers on the $1.2-billion field production system contract for its side of the field, which is being executed by Ralph M. Parsons Company and a batch of subcontractors. On the BP side are two more 500-person camps for the contractors building the field systems there—Brown & Root, Inc., and Locher Company, units of Halliburton Company.

Despite whatever comforts of home they provide, the installations at Prudhoe Bay are basically hardware—no more nor less than what's needed to operate what may be the biggest single United States oil field ever brought into production. The structures dotting the polar plains of the North Slope are all the more remarkable when it's remembered that every nut, bolt, pipe, and compressor was brought to the Arctic shore from thousands of miles away. The main means of carrying out this ten-year logistical extravaganza have been the annual barge flotillas coming around Alaska from the Pacific Northwest and other places even more distant. For the Prudhoe Bay field itself, these began in earnest in 1970 when 187,000 tons of materials were barged up to the Slope.

They've continued each year—the 1976 shipment, after a relatively uneventful six-week voyage, was another 75,000 tons, including 102 of ARCO's field modules and 33 modules for BP Alaska.

But no one at the slope had yet forgotten the 1975 barge episode. For the two field superintendents in the operating halves of the field, BP's Charlie Wark and ARCO's Landon Kelly, the real work only started when the last of the 1975 barges were frozen into the offshore ice a mile and a half from Prudhoe Bay's docks that October. The rest of that year was spent getting permits for yet another quickly designed piece of Arctic engineering—a 5,300-foot gravel causeway, 220 feet wide at the bottom to distribute the weight of the heavy shipments that had to be brought off the barges over the unfrozen silted floor of the bay.

The causeway was built the following winter, by cutting away the ice along the path of its route in eight-ton blocks, then moving in quickly to structure the causeway with load after load of gravel. The modules, of up to 1,300 tons apiece, had been welded to tracks on the barges for the trip up, and the barges also carried 800-ton crawlers for off-loading their massive pieces of cargo. The crawlers were another legacy of the United States space program—similar machines had been used to move the towering Saturn rockets from their assembly buildings to the launching pads. Over the rest of the winter, the loads on each barge were taken off along spurs onto the new causeway. The last module had crawled ponderously ashore by spring—in time to keep the construction of the field production systems on schedule.

The 150 development wells at Prudhoe Bay will supply the Alaska pipeline's needs at its 1.2-million-barrel daily throughput rate. More wells will exceed this somewhat, allowing up to 1.5 million barrels a day if the pipeline begins building up to its maximum designed capacity of 2 million daily barrels a few years from now. Before that full rate can be reached, though, yet more oil has to be found at Prudhoe Bay.

Side by side with the drilling of development wells into their proven reserves on the North Slope, the oil companies have kept on drilling exploratory wells to see what more they can find. Some of this drilling has been in the same acreage as the initial Prudhoe Bay discovery well and the later wells that confirmed it, where oil-bearing sands at other depths had already been found but not fully proven out. More drilling has been done away from the proven acreage, on leases acquired in the $900-million Alaska lease sale of 1969—and, for that matter, on leases from the several earlier sales that had never been fully probed.

Oil and gas are found in the earth in the more porous pockets that may occur between layered strata of rock dating back through millions of years of geologic time. These strata have been identified and named according to a standard worldwide classification system. In a possible new oil area, geologists and geophysicists first look for rock outcroppings on the ground surface that might give clues to what lies below. Then, by means of small detonations, they create miniature seismic waves that are recorded and analyzed for what they may show of the depths and structures of underlying rock strata. The art and science of geophysics is in judging how the strata are arranged in relation to each other, and what the likelihood may be of hydrocarbon-bearing pockets between them. Then comes the expensive step—still against great odds of success—of drilling an exploratory well.

The first strike at the North Slope was made in the Sadlerochit sands that are part of what geologists have named the Jurassic-Triassic formation. At Prudhoe Bay, this is found between 6,000 feet and 9,500 feet below the surface. That well and later ones also showed oil and gas in a deeper formation, the Lisburne limestone that lies between 9,000 feet and 13,000 feet. The Lisburne formation extended under most of the Sadlerochit and also somewhat to the east of it. Atlantic Richfield has drilled two successful wells into this formation, and is continuing these explorations.

Yet another promising oil zone at Prudhoe Bay lies higher than either the Sadlerochit or Lisburne pools, and to the west of them. This is in the Upper and Lower Cretaceous formations, found variously from almost sea level down to 6,000 feet. In the Lower Cretaceous rock, ARCO, SOHIO, Mobil, Phillips, and others have drilled successful wells into the Kuparuk River sands, and more exploration is being done to define whether this pool contains commercial quantities of oil. A similar find was made in the Sak River sands of the Upper Cretaceous formation, again west of the main Prudhoe field. But this oil so far has proven to be a heavier crude that may not be commercially worth producing. Both ARCO and SOHIO have also drilled exploratory wells on some small offshore islands a few miles from the shoreline at Prudhoe Bay, and these programs are continuing.

Nor is Prudhoe Bay any longer the only part of the North Slope that is being drilled. Over the winter of 1976–77, the U.S. Navy again began an active exploration program in Naval Petroleum Reserve No.4, where small oil discoveries after World War II first lured British Petroleum and Sinclair Oil to lease and drill on the slope. Pet 4's 23 million acres cover the northwest of Alaska from the Colville River west to Icy Cape and south to the continental divide of the Brooks Range—Prudhoe Bay itself could be lost in one corner of the naval reserve's vastness. The Navy has contracted with Husky Oil Company, the U.S. unit of Canada's Husky Oil Ltd., to do 10,000 square miles of seismic mapping and drill twenty-six exploratory wells over the next five years. The first seven wells were drilled in the winter of 1976–77.

For the future, the next big exploration effort in northern Alaska may come in the offshore islands of the Beaufort Sea, a large water province north and east of Prudhoe Bay and extending over into Canadian waters north of the Mackenzie River Delta.Canadian companies have been exploring there for the past two years, so far without any announced success, under drilling conditions that are compounded in difficulty by being both Arctic and offshore. On several occasions since 1974, the state of Alaska has scheduled and then postponed a lease sale in the Beaufort Sea, and it remains one of the largest known sedimentary basins under state control that has not yet been probed.

With the soaring costs of the Alaska pipeline—and of developing the Prudhoe Bay field to feed it—oil companies may be just as pleased that another major lease sale in the north has been held off. On the other hand, Alaska's part of the Beaufort Sea has got to have a tantalizing allure for petroleum seekers simply because it does lie so close to the towering bonanza of Prudhoe Bay itself.

17.
1977 ...Turning
the TAP

As 1977 began in Alaska, there was quite a different look to the spectacular undertaking that had dominated the state's life for so many years. All of Alyeska's efforts since 1968—all of the tens of millions of man-hours of design and construction planning and just plain hard work—were coming together at last. This was the year the Alaska pipeline "project" became the Alaska pipeline.

The project was finishing on schedule. In November 1973, with its permits finally in sight, Alyeska had projected a completed pipeline ready to take on oil by mid-1977. Despite everything, that schedule was being met. "Everything" included Alaskan weather: early winters and late springs, too much snow and not enough snow, drenching rains and high winds. It included Alaskan terrain: permafrost and ice lenses, collapsing rock and mud slides, roaring rivers and melting glaciers and *aufeis*, vertical canyon walls and steeply pitched mountain passes. It included countless engineering adaptations: special burial structures, several kinds of bridges, new wall-building techniques, air-cushioned river craft and helicopter airlifts. It included late and erratic deliveries of critical materials. It included labor flare-ups and fallible contractors. It included cor-

195

rupted welding records and an extensive remedial program to re-
certify work already done to strict specifications. It included costly
and time-consuming government surveillance at all levels—the
building of unprecedented redundancies into every part of the
pipeline system. Doggedly, the schedule set in 1973 had survived
and overcome all of this.

Now, from Prudhoe Bay to Valdez, several things were going
on at once. The last bits and pieces of construction were being
completed. The route was being cleaned up from end to end, while
this year's restoration and other environmental work got under
way. Demobilization had already begun—the unmistakable signs
of a mammoth construction project winding down. And large parts
of Alyeska Pipeline Service Company were already looking like the
operating company Alyeska was about to become. Operating personnel were taking over the mainline sections, pump stations, and
terminal facilities from the construction contractors as each piece of
the system was finished. Preparations were all but complete for
1977's long and intricate process of starting up the $7.7-billion oil
transport and shipping system.

Much of the remaining work along the pipeline route had
been let to a new set of contractors. All of the final mainline construction tasks went to Associated-Green, the execution contractor
for the 143 miles of Section Four that ran north from the Yukon
River into the southern foothills of the Brooks Range. This joint
venture group had brought its section along in good time during
the two full years of mainline work, including an added 18 miles
south of Dietrich Camp that had been transferred from Section
Five as difficulties arose there. Associated-Green now took on the
last 160 miles of hydrostatic testing, the last forty-five miles of
above-ground insulation, the last thirty-three remedial welds, and
other final construction touches up and down the line.

The contract also included civil construction work north of the
Yukon. Part of that work was building sixty-two river training structures along the route, and moving about half a million cubic yards
of rip-rap and overfill material into place at selected points on the
line. It also meant road and camp maintenance, cleanup of the
entire system and its surroundings, revegetation work, and the
tasks of demobilization. A second contract for the civil work south
of the Yukon went to Alaska International Construction, Inc.,
which had been doing haul road maintenance in the north.

The other big piece of pipeline work in 1977 was the last

seventy-seven miles of the small-diameter gas fuel line from Prudhoe Bay to Pump Station 4 just south of Galbraith Lake. This was awarded to Houston Contracting Company. The job was started in January, with completion of the combination ten-inch and eight-inch line set for spring.

For the first time as a construction season opened, spring was coming early to Alaska this year. "Florida Gets Snow, While Alaska Basks," headlined *The New York Times*. In the worst winter of the century, the entire eastern half of the United States was nearly frozen solid. In Anchorage, though, hockey games couldn't be played because the ice rinks were melting. Fairbanksans were running around in sweaters. Even at Prudhoe Bay, said the *Times*, "the temperature was a balmy zero." A huge military training maneuver had been held in central Alaska each winter because it was the coldest place to be found. After this year, the Arctic exercises were shifted to Watertown, New York—its weather early in 1977 had been more Arctic than the Arctic. Up and down the pipeline route, in January, work crews were rushed out into the field to take advantage of every warm day. This was the first big weather break the project had gotten, and it had come perhaps when it was most welcome—at the end.

The pace of the construction work left to be done in 1977 was just as urgent. But there was not as much of it, and this was reflected in vastly reduced work force levels as the year went on. Total manning levels on the project, including craft labor and administrative jobs, had peaked out above 21,000 both in 1975 and in 1976. They were not to be much above half that number in 1977. By mid-April of 1976, more than 18,000 people had already been at work on the project. By the same time in 1977, less than 11,000 were employed. A total of 19,000 craft workers had been on the project at its peak in 1975, and 18,000 in 1976. The peak in 1977 was to be about 6,500. Craft workers had logged more than 26 million man-hours in 1975, and only somewhat less than that in 1976. The wrap-up work in 1977 was expected to take about 3.5 million man-hours of craft labor.

Alyeska's thirty construction camps—an extra one had been added at Sourdough early in 1976—were gradually being shut down. Seven camps had been closed last November: Prospect Creek and Old Man in the Brooks, the Livengood camp near Fairbanks, and the camps at Delta, Isabel Pass, Sourdough, and Tonsina in the southern sections. Six more camps closed over the winter. The shut-down camps were mothballed until their fate, along with most of the rest of Alyeska's camps, was decided. That fate was not clear as yet. Alyeska might keep a few of its camp facilities for use by maintenance crews after the pipeline was operating. But the stipulations required that most of the camps, at some point after 1977, had to be dismantled and taken out of the landscape. Two of the three proposed projects to bring out Prudhoe Bay's natural gas would use all or part of the same route as the oil pipeline, and so could use the same camps. But that government decision was not due until late in 1977. Meanwhile, at least twenty of the facilities were up for sale.

Alyeska was going into the surplus business in a large way. "It will be," the *New York Times* said, "one of the biggest going-out-of-business sales in history." In January the company listed more than 20,000 pieces of equipment for sale, with an original purchase value of about $800 million. Like everything else about the project, some of this was big equipment, and there was lots of it: 240 cranes, 119 backhoes, 719 bulldozers, pipelayers, and loaders, 1,340 generators, 1,357 trucks of all types, 3,315 vehicles ranging from station wagons to buses and ambulances, 1,637 welding machines, and on and on. There were also large batches of spare parts, lumber, office supplies, and thousands of pieces of Arctic work clothing. It was noted that the list included 1,500 gas-heated outhouses at an original unit cost of $10,000 each. Anyone who had worked along the route in winter knew these were not luxury items.

The surplus sale, which would probably take a couple of years to complete, was not just a matter of cleaning up Alaska after the end of the project. The equipment had had hard use. But it had also been well maintained to keep it operating, and could still be expected to fetch a decent price even as used goods. The several hundred million dollars that ought to result from the sell-off were to be a net deduction against the final total of the project cost. That cost in turn would go into determining the rate of return that Alyeska's owner companies would be granted on the pipeline. That would help determine the pipeline's taxes and tariffs. The transport costs of North Slope oil, in part reflected in the pipeline's costs, would be a factor in setting wellhead prices at Prudhoe Bay. And this would affect the state of Alaska's income in years to come from its 20 percent share of the oil price in royalties and severance taxes on Prudhoe Bay's production.

But first there was start-up. In April, Alyeska filed notice of intent to begin filling the line between June 20 and July 1. That long-awaited "oil in" date was at last near at hand. William Darch had been in Anchorage almost a year now as Alyeska's president and prospective head of the operating company. Studies and tests toward the first operations had gone on all fall and winter. Now the plans were in place, and the timing and methods of start-up could be spelled out. As Bill Darch had said the previous summer, nothing about putting a pipeline in Alaska into operation was going to be simple. He talked more now about some of the technical confrontations of start-up.

These began with hot oil and a cold pipeline. "We've got two interactions here," Darch explained. "The oil tends to cool down . . . and the pipeline warms up. Ultimately, we get a balance between the two." The temperature spread at first was gaping. Prudhoe Bay's oil was coming from its wellheads at 160°. Gas extraction and other processing by the producers in the field cooled it to 140° by the time it was ready to enter the pipeline. At start-up, though, the pipeline itself was at an average temperature of about 20°. The first contact between the two was like ladling hot soup into a cold icebox dish.

So the need to do things differently in Alaska came right away. "The conventional method of starting up a pipeline," said Darch, "is to fill the pipeline with water, then place a separator 'pig' in the line, then introduce the oil after that . . . and move the crude oil through the line. The object of putting water into the line is to purge air from the system . . . air, of course, contains oxygen, and in mixing oxygen and hydrocarbons you always have potential safety problems. But with the coldness of our particular line, if we put water in it, it's likely to freeze and we could end up with the longest Popsicle in the world."

The answer to that in the Arctic was to substitute nitrogen, an inert gas that can't support combustion, for the water. On June 20, Alyeska shut the first gate valve in the mainline, 18.2 miles south of Pump Station 1, and filled that part of the pipeline with nitrogen to a pressure of eighty-five pounds per square inch. With oil ready to go in, that valve was then opened and the nitrogen expanded to atmospheric pressure, bringing it down the line another ninety miles and purging the air out ahead of it as it went. Now the separator pig went into the line, and then the crude oil. North Slope oil at last was on its way to Valdez.

That first trip was a slow one and there were serious difficulties. Oil was introduced to the line at an initial rate of 300,000 barrels a day, to be boosted up to 600,000 a day in about thirty days, and 1.2 million barrels a day by the spring of 1978. The pipeline alone when filled held 9.2 million barrels of oil, or almost an eight-day supply of its normal throughput. At the initial charge rate, it was to take thirty-one days for the first oil to reach Valdez.

The oil was moving at just over one mile an hour. And, within the chilled walls of the steel pipe, it was very cold and sluggish oil. "As oil cools down, it becomes thicker," explained Darch. "Our oil at first had a viscosity very similar to asphalt." Its first destination, of course, was Atigun Pass, the line's highest point at 4,790 feet. Three pump stations on the north side of the Brooks were pushing it toward the summit of the pass. Even with the initial head of pressure needed to move this first chilled oil over Atigun, though, the hydraulics of the line were still well within the maximum design pressure limits.

On the downslope of the Brooks, start-up raised another technical challenge. It was necessary to make sure that the oil, behind its pig, always moved in a continuous thread. Coming over Atigun Pass, that thread could "break"—and a portion of the oil, with the pig, would go charging downhill. "One has the analogy of a runaway locomotive in terms of weight and momentum with a slug of oil running down that freely . . . it's a very sizable chuck of momentum," said Darch. The risk here was that the unchecked oil, as it reached bends in the pipe along the route—especially in the relatively unrestrained above-ground pipeline—would pack enough wallop to damage the pipe. The solution here was to close another gate valve on the downhill slope and install a pressure controller on the bypass pipe around it. This was used to maintain a back pressure on the nitrogen, and therefore on the pig, to keep the oil from the north moving downhill just as slowly as it had come up the pass from Prudhoe Bay. During start-up, until the line was full, this procedure was repeated over every pass all the way to Valdez.

Then there was the wax. This had radically changed the project's design concept from a chilled oil pipeline to a hot oil line very early on. "All crude oil, as you cool it, does precipitate some wax," said Darch. "There are two opposing effects that take place. As you cool the oil down, the solubility of the wax decreases, and more of it settles out of solution. But the countermechanism is the viscosity of the oil, which also increases as the temperature drops . . . and as the viscosity increases, the rate of settling of the wax decreases." Alyeska had spent the winter pumping oil through the 8,000 feet of four-inch, six-inch, and eight-inch lines it had built at the North Slope. From these tests, it could extrapolate up to a firm idea of how much wax precipitation to expect in the 48-inch pipeline, and how to control it. The answer: more pigs, of a type that would swirl the wax through the oil and keep it in solution.

All through the start-up period, the confrontation of hot oil versus cold steel required a careful watch on the shifting of the pipeline itself. "We have 800-plus miles of continuous steel pipe, and as that changes temperature, it's going to grow in length," said Darch. Several design elements had anticipated this: the trapezoidal zigzags in the elevated pipeline, and the careful alignment of the pipe, based on computer calculations, in its sliding shoe on the crossbar between each pair of VSMs. It was not beyond happening, Alyeska knew, that excessive pipe movement would require shifting the position of some VSM pairs to compensate for more sideways movement than expected. "It's a very remote possibility," Darch said, "but it might even turn out, in some locations, that we've got too much pipeline, and will have to take a few feet out . . . or even put an extra portion in, if we haven't got enough pipeline somewhere."

Each of these possibilities had been thrashed out at a series of "what if " sessions held to conjure up every conceivable event that might occur during start-up. Other touchy points along the route were the several suspension bridges at river crossings. As the first oil came across a bridge, it would unbalance the initial loading of the structure. Alyeska had determined, through more tests, that at the initial throughput and slow pace of the oil, each bridge was a safe situation.

From design through start-up planning, repeated calculations and computer runs had simulated every detail of operating the pipeline. The system was instrumented, of course, from one end to the other to tell its operators almost instantly what was taking place anywhere on the route. Even so, Alyeska almost literally held hands with its first oil in the line over the entire 800 miles to Valdez. "Never mind how well you've done your precommissioning of your measurement system. Typically, there is going to be a chance of a failure," said Darch. "There are something like a million electronic components in our leak detection system, and it would be unwise to rely entirely on our instrumentation until we've gone through this initial start-up phase."

As the oil front moved down from Prudhoe Bay, crawling along at its mile-per-hour rate, a ground party walked along the pipeline beside it. A second party followed twenty-four hours later, and a third party twenty-four hours after that. Their first job, obviously, was to watch for oil leaks. They also carried gauges to check clearances between the shifting pipe and the VSM uprights. And they also checked to see that the VSMs were taking the load of the filling line without distortion. Even after the oil front had passed, more ground parties continued walking north and south along each short portion of the line for several weeks more to keep up the vigil.

The main goal during start-up was to bring the pipeline system into thermal equilibrium as the warm oil and cold pipe interacted. "Thermal equilibrium simply means that the oil is coming out at consistently the same temperature," Darch said. "We expected that, at a throughput of 600,000 barrels a day, it would take up to 120 days of continuous operations to reach a thermal balance." This doesn't mean that the oil emerging at Valdez is as warm as it was—140°—when it left Pump Station 1. By design, that exact balance won't be reached until the pipeline is at its maximum capacity. The first oil at Valdez, in fact, arrived at barely 45°, gradually rising in temperature from then on as the entire line warmed up from the continuous flow.

Perfect thermal balance—140° versus 140°—means the pipeline system is generating as much heat as it is losing. Heat is lost from the initial action of oil against steel, and again as it cools down somewhat between pump stations. Heat is added to the system, at the same time, by the initial temperature of the oil, by friction as it moves along the pipe, and by the pumping action itself at the stations. More of the heat loss is offset as throughput increases, and that will occur over the early years of the pipeline system's operations. The coldest the oil should ever be was on the first day of start-up. "Our concept," said Bill Darch, "is that we start up this pipeline once . . . and that's now. It will stay onstream then forever . . . 'forever' being the life of the oil field."

Start-up went smoothly until the oil reached Pump Station 8, almost 500 miles down the line. There, nitrogen being put into the pipeline was not heated, and extreme low temperatures cracked a pipe bend. Then on July 8, workers were replacing a strainer at a shut-down pump, when, through an error, the pump sprayed oil from the open strainer into the pumphouse. The fumes ignited, and fire heavily damaged the building, killing one worker and injuring five.

The station was bypassed and oil flowed on to the terminal by the end of July. The Alaska pipeline system had one more "first" task to perform. This was to load and dispatch the first tanker of Alaskan oil from Valdez.

A tanker had been at Valdez as early as April. The *S.S. ARCO Fairbanks* arrived at Bligh Island just south of Valdez Arm on April 6. At 120,000 deadweight tons, it was the largest and newest United States flag tanker in the Atlantic Richfield fleet. Boarding the ship at Bligh Island were more than forty ship captains, state of Alaska harbor pilots, and Coast Guard personnel. Also aboard were representatives from the marine departments of ARCO, Exxon, Mobil, SOHIO, and several other oil and shipping companies. Half-ballasted to a draft of thirty-five feet, the *Fairbanks* had sailed up to Alaska from ARCO's refinery at Cherry Point, Washington. It was to spend the next month as a training ship for the super-port that would go into operation at Valdez this summer.

By the end of that month, the seafarers would know the waters of Prince William Sound and the Valdez harbor well. For the first two weeks, the *Fairbanks* made some fourteen round trips across the sound and through the outer parts of the Valdez Narrows. This was to qualify the tanker captains for pilots' licenses in the sound, and to show the Alaskan pilots the capabilities of the large ships they would be guiding in and out of Valdez. Navigational equipment was demonstrated and tested. And the Coast Guard, which had established a permanent installation in the Port of Valdez as the time for tanker operations came near, was there to study the terminal's port navigation and communications systems.

The second two weeks of the training cycle brought the *Fairbanks* into Valdez for another fourteen daily trips in and out of the terminal facilities. The tanker was ballasted to its fully loaded draft of fifty-two feet. Tug maneuvers were practiced, and there were repeated dockings and undockings at the terminal's two completed berths. The Valdez shipping plans called for one-way tanker traffic in and out of the harbor, on an initial frequency of six to eight ships a week. The tanker channel is 2,700 feet wide, running past Middle Rock, an undersea pinnacle that jutted thirty feet out of the water in the center of Valdez Arm and was already well beaconed by the Coast Guard.

The April trials were a dress rehearsal for the main event itself. Then, on August 1, 1977, the ARCO *Juneau* sailed with oil from Valdez. The departure of that first ship wasn't an anticlimax at all. It was, after nine years, simply getting down to the business that the Alaska pipeline project was all about. The first U.S. oil from the Arctic, hard-won and long held in reserve, was heading south into the needful mainstream of American energy marketplaces.

18. The Alaska Pipeline... Tomorrow's Frontier

Framed on the wall of Charlie Spahr's office in the Midland Building in downtown Cleveland is a retired certificate for 10,000 shares of stock in The Standard Oil Company. Beside it is a copy of the corporate charter, dated January 10, 1870. There is also a photograph of John D. Rockefeller, the company's founder-president, who used Ohio Standard as the core of what would become the Standard Oil trust—controlling dozens of other oil companies across the United States. These, when the trust was dissolved into thirty-four independent companies by the U.S. Supreme Court in 1911, would be the seeds of much of the modern U.S. oil industry.

Chief executive since 1959 of SOHIO—as The Standard Oil Company (Ohio) came to be called—Charles E. Spahr talks of having known, in their later years, some of the Rockefeller associates. "There are not many industries," he says, "that you can almost group in a lifetime."

Few companies, even in an industry whose growth has all taken place in not much more than a century, have seen the explosive change that the years of Charlie Spahr's tenure have brought to SOHIO. The Standard Oil trust breakup was mainly geographic, leaving the separate companies to go forward where they stood with what they had. In SOHIO's case, that was a small refinery and a one-state marketing territory—no crude oil of its own, no pipelines to bring the oil it purchased to its plant. For more than fifty years, SOHIO would work its way uphill, building its own exploration and production strength and its own pipelines, and pressing its markets beyond the borders of the state of Ohio. By the 1960s, SOHIO was a major regional refiner and marketer and had expanded nationally into chemicals and plastics. But it was still an unbalanced middle-sizer among the oil giants—badly short of its own crude-oil supplies.

That all changed sharply in late 1969, with the amalgamation of SOHIO and British Petroleum—and Alaskan oil. Corporately, as North Slope production begins swirling down the Alaska pipeline, SOHIO becomes part of the worldwide BP Group of companies. Yet it remains largely an autonomous U.S. company with its own shareholders, management, and operations. Those operations, though, through BP Oil, Inc., as a U.S. unit of SOHIO, now include refining in the Northeast and Southwest, marketing in the thirteen eastern seaboard states. And, of powerful importance, SOHIO holds a whopping half share in the proven reserves at Prudhoe Bay.

SOHIO is thus totally turned around. "The North Slope and the pipeline, now that they're completed, make us a completely integrated oil company," says Spahr. "In fact, we'll be overintegrated on the production side . . . that's kind of a different situation for us, believe me." Its new and growing crude supplies place SOHIO in the top ranks of U.S. oil companies in terms of domestic production. And that means opportunities on the "downstream" end of the chain of petroleum activities: expanded and more efficient refining and marketing, and growth also in chemicals of all sorts with an assured source of feedstocks now from Alaska. "No one has yet found a substitute for hydrocarbon raw materials in plastics and fertilizers," says Spahr as an example. "The hydrocarbon," he adds, "is such a versatile product beyond fuels."

For British Petroleum, too, the linkup with SOHIO in the United States is part of a dramatically altered corporate profile for the future. BP has been long on success where SOHIO was short—in exploration and production. BP is one of the most skillful finders of crude oil in the business. Downstream around the world, though, BP has not been as strong in marketing its oil. It had long sought an entry into the large U.S. oil markets. BP might simply have acquired an American oil company. Instead, it traded for a majority investment in one. Its bargaining chip was an inpressive one: more than 4 billion barrels of Alaskan crude oil—the west side of the Prudhoe Bay oil field.

What has Alaska and the pipeline gained for BP? "From a position where we had very little in the U.S., Alaskan oil made us second only to Exxon . . . with the second largest U.S. reserve position," says Montague M. Pennell, deputy chairman of British Petroleum Ltd. in London. A petroleum engineer, Pennell was with BP North America in New York when the company was casting up its first North Slope drilling plans. "The other international oil companies have always had large stakes in the United States . . . which has been and is the largest market. We hadn't," Pennell says. "So, because of what underlies those leases on the North Slope, we are now a large factor in that market, on the American scene."

BP's stake in the U.S. is a proportion of SOHIO's stock that rises from 25 percent to 54 percent as the BP/SOHIO share of Prudhoe Bay production increases to 600,000 barrels a day. BP, in effect, has joined SOHIO's 40,000 other shareholders, most of them in the U.S., who will still own the remaining 46 percent of the company when the BP portion reaches its maxium. "Whatever we get from the North Slope now will be in the form of dividends," Monty Pennell says. "On the other hand, we will have a controlling interest in a company that should have a very bright future, even though it will be a more or less autonomous part of the BP Group."

Atlantic Richfield, although it was taking a different route to get there, was bound for the same destination as SOHIO. The merging of Atlantic Refining, Richfield, and Sinclair into the new ARCO in the 1960s created a larger and more diversified company—but still an unbalanced one. ARCO's oil production was only somewhat more than half of its refining, marketing, and chemical feedstock needs—the rest was imported from overseas. That problem is virtually erased by Alaskan oil. ARCO's share of the proven reserves at Prudhoe Bay is estimated at nearly 2 billion barrels—now almost half of the company's total worldwide reserves. So from a single source, ARCO also reaches self-sufficiency in its own oil supplies.

"Our basic objective in this company is to be in balance, in the United States, with our own sources and our own outlets," says Thornton F. Bradshaw, president of Atlantic Richfield in Los Angeles. "Industry-wide, the cheap oil is gone, and there will be increasingly high costs for sources from now on. Now consider this extraordinary opportunity in Alaska. It puts this company in a differential position. As we see that increasingly expensive oil, we don't have to participate . . . *we* have options."

Houston is the headquarters of Exxon Company USA, the domestic affiliate of Exxon Corporation. Exxon is the largest industrial company in the world. Its current sales and assets alone almost equal the combined sales and assets of all of the other U.S. owners of the Alaska pipeline. Its worldwide oil and chemical activities stretch from the Middle East and Europe and the North Sea around through the Pacific to Latin America, Canada, and the United States. Exxon USA itself, standing alone, would be bigger than all but a handful of American companies. It has, in recent years, contributed almost half of its parent corporation's total earnings.

Now North Slope oil has come into production. It's been popular to suggest that Alaskan oil might not be so important, relatively, to a company as large as Exxon. But Exxon USA's stake in the Prudhoe Bay oil field is roughly 2 billion barrels. That is about 40 percent of the company's total domestic reserves. With oil discoveries and production levels in the Lower 48 tailing off throughout the industry, those Alaskan reserves are a big part of Exxon's future. "Let me assure you that's the case," says George E. Uthlaut, executive assistant to Exxon USA's president, Randall Meyer, and the company's representative on the Alyeska construction committee. "Prudhoe Bay and the pipeline will be giving us 240,000 to 250,000 barrels a day by next year, and we don't look at that in a casual fashion . . . it's important to us."

Important, indeed. Alaskan production initially will be more than 20 percent of Exxon's total domestic output. Optimistically, some Wall Street analysts have projected, the North Slope could account for almost half of the company's U.S. oil production in the 1980s. Exxon USA produced an average of 940,000 barrels of oil a day in 1976, a decline of 3 percent from the year before, and maintaining a downward trend that goes back to 1971. The quarter of a million barrels daily from the North Slope will sharply reverse that trend in the years ahead. And further successes on the Slope could improve the situation even more—for Exxon, and for all of the producers there.

The other Alyeska owner companies—Mobil, Phillips, Union Oil, and Amerada Hess—can also count on percentages of the pipeline throughput that are roughly proportionate to their ownership shares in the pipeline system. Those are smaller shares, to be sure—ranging from 5 percent down to 1.5 percent. And many of the companies have partners in the oil field. But measured against the formidable volumes of oil waiting to be pipelined from Prudhoe Bay over the life of the field, even the lesser shares take on a dimension that, by oil industry standards, is pretty husky for a single source of crude oil. If Prudhoe Bay is reckoned to hold proven reserves now of almost 10 billion barrels of oil, even a 1 percent share of that is about 100 million barrels. That's a lot more than a drop in anyone's oil bucket.

Completion of the Alaska pipeline, says Exxon's George Uthlaut, "means an opportunity for us to recoup some of the massive investments we've put into Alaska." For all of the Alyeska companies, the $7.7-billion cost of the pipeline system itself is only a part of the overall price of their Alaskan ventures—and not even an overwhelmingly large part. Each of the companies, according to its share of the project, has had to fund its part of the costs through a succession of public and private bond offerings during the years of construction. The biggest of these was a $1.75-billion issue arranged for SOHIO and British Petroleum by Morgan Stanley & Company in November 1975. It was sold to seventy-six insurance companies and other financial institutions—the largest private placement of notes in corporate financing history.

But the interest costs of these issues—about $1.4 billion spread among all of the companies—must also be added to the project's total costs. Then there is the cost of developing the Prudhoe Bay oil field itself, including all of its production and gathering facilities. That has been roughly another $6 billion, plus interest costs. Many of the companies are also financing construction of new tankers to carry oil from Valdez, and those costs could be considered a part of the companies' Alaskan outlays. Add it all up, and the cost of developing North Slope oil and bringing it to market over the past ten years represents an oil industry investment of something more than $15 billion. Not a cent of that was earned back until Prudhoe Bay's oil began flowing this summer.

In May 1977, just before the oil did begin flowing, Ed Patton visited the Overseas Press Club in New York. The Alyeska chairman was also talking in billion-dollar figures, but some of them were on the plus side. "By 1978," Patton told his listeners, "the Alaska pipeline-Prudhoe Bay venture will be generating for the United States a favorable balance of payments effect of $15 million a day, or about $5.5 billion for a year of operation at 1.2 million barrels a day. This favorable effect would have about canceled out the U.S. balance of payments deficit experienced in 1976. But unfortunately, it will probably not cancel out the deficits we are currently experiencing."

The week before, in fact, the U.S. Department of Commerce had announced some grim trade figures for the first three months of 1977. The excess of imports over exports in the period was $5.92 billion, which was just $50 million more than the entire trade deficit for all of 1976. Oil imports in March alone were $4.06 billion, as U.S. fuel oil stocks were being built up again after the harsh winter. The March trade deficit was more than twice as high as any single monthly deficit ever recorded until 1977. And it was all oil. Without those oil imports, there would have been a trade surplus of more than $1.5 billion.

"North Slope oil will contribute more to reduce U.S. dependence on foreign oil than any other action taken in this decade," says Randall Meyer of Exxon. True enough, Ed Patton told the Overseas Press Club: "The availability of 1.2 million barrels a day from Prudhoe increases domestic crude supplies by 15 percent." But, added Patton: "To look at it another way, we still need about five more Prudhoe Bay developments to rid ourselves of dependency on foreign oil."

Where are these developments likely to take place? The clue to this is in the story of the last ten years in Alaska. Ed Patton talked of the superlatives of North Slope oil and the pipeline project he had guided to completion. "The venture brings to market the product of the largest oil field yet discovered in North America. The Prudhoe Bay field now represents about 30 percent of the proven oil reserves of the United States," Patton said. And it had another important distinction. "When this oil reaches the marketplace," said Patton, "it will be the first frontier oil to have done so. And, as we all know, most of our future oil and gas supplies must come from the so-called frontier areas."

The Alaska pipeline has been a frontier venture in every possible sense of the word. The project has spanned the gap between a time when the world still thought its supplies of energy were inexhaustible—and today, when the world knows at last that they are not. Paradoxically, it has taken one of the largest oil discoveries ever made—and one of the longest and most expensive construction ventures in history—to prove the point. Finding and producing another five Prudhoe Bays had better not take so long. For United States energy needs, there is not that kind of time left.

Technically, the Alaska pipeline has been its own frontier. There will be, unquestionably, other large and complex energy projects in the Arctic and at the other distant edges of civilization. Of its kind, this was the first. For Alyeska's managers and engineers and builders, coming to Alaska from all around their industry and outside it, there has never been a project like this one. It is hard for many of them to imagine what next they might do in their careers to match the scope and challenges of this assignment. The pipeline's technology is the state of the art in every way and, in many instances, well beyond it. It will be a textbook for any but the most conventional pipelines for decades to come.

The frontiers of Alaska itself have been pushed forward by decades. To say that the Alaskan frontier has been tamed or spoiled by the pipeline project is simply to say that you do not understand Alaska. A rugged landmass of hundreds of thousands of square miles on the borderline of the Arctic will surely be affected by the coming of a large industrial development. But it will hardly be overwhelmed by it. Alaska has been affected. The "boomers" have come again, as they have so many times in Alaska's past. They've come by the tens of thousands, and most of them by now have gone away again, taking their noise and their money with them. This time, though, it isn't Alaska's wealth that they've carried away—its gold or its copper, or its furs. And some of the newcomers, as before, have stayed—liking what they saw of Alaskan life away from the furor of the construction project, and wanting to be a more permanent part of it. This, too, has happened in the past, and the state has been enriched by it, for newcomers like these have often become the best kind of Alaskans.

This time, too, Alaska has been enriched by real wealth. Alaska's true financial worth is in its resources. Its timber and fishing industries have been early versions of this. Now an Alaskan resource is being coined into the world's leading currency: oil money. This is not federal money pumped in from Outside by an absentee landlord. It is coming out of Alaska's own earth: $1 billion a year or more, by the 1980s, that will give the state the economic base that its people have earned—a reward for their long years of having to do it all themselves, and a reward for giving the world what it wants. The nations of OPEC, also stewards over unbounded oil, do not do this so generously.

Alaskan natives have surely been brought forward by decades, if not centuries. Their social struggles are not over at all. But they now have the best of both the old and new worlds of Alaska: A genuine title to the land that was always theirs anyway; and a handsome portion of the oil money to use in fighting the poverty and backwardness of their history of second-class citizenship. They have already made a beginning on the industrial skills and financial structures and modern businesses that will bring them forward even farther. This is a remarkable experiment in social engineering, and Alaska is the scene of it.

It has also been the scene of the environmental confrontation of its age. Nowhere else has the basic dilemma between wilderness preservation and resource development been fought out so long, so bitterly, so expensively, and so instructively. Alaska has frontier oil, and other resources, that the world feels it needs. Can they be extracted without destroying the environment in which they're found? The oil field on the North Slope and the pipeline leading from it have been another remarkable experiment—this time in the coexistence of wilderness and development, and the ability of Arctic and sub-Arctic regions to play an undamaged host to both. A lot has already been proven on both sides of the equation: the ability of industry to do its work responsibly, and the ability of the environment to stand up to the pressure.

A flight out of Alaska, in 1977, is an airborne epilogue to the Alaska pipeline project. The pipeline workers are going home. Anchorage International Airport, for years, has been an incoming conduit of trunks and duffel bags and toolboxes. Now the baggage of the pipeline, and the lusty Texas and Oklahoma accents of the pipeline's men and women, are going the other way: Outside.

On the packed airliner, the knowing flight attendants make an early circuit with coffee and sweet rolls that scarcely dent the enthusiasm from that last round of farewell parties in the airport bar. A welder in a denim leisure suit is straining against his seat belt, breaking up with hilarity, spilling his coffee, banging his friend on the shoulder. "Hey, man," he is laughing, "you got strawberry jam all over your mustache."

The plane circles up slowly over the sprawling night lights of Anchorage and heads away down the Cook Inlet, bound for the Lower 48. Alaska recedes, and with it the colors of hundreds of sunrises and sunsets over the Brooks Range, across the Copper River Valley, above Valdez Arm and the Chugach. Alaska, in these years, has been the place where the world has come to correct all of its mistakes at once. Perhaps to an amazing degree, it has succeeded. Alaska and the Alaska pipeline have been energy's laboratory. And, in a great many ways, a laboratory of today's society as well.

Bibliography

Adelman, M.A., editor, *Alaskan Oil—Costs and Supply*. New York: Praeger, 1971.

Agreement and Grant of Right-of-Way for Trans-Alaska Pipeline. Washington, D.C.: U.S. Government Printing Office, 1974.

Alyeska Pipeline Service Company, *Summary, Project Description of the Trans-Alaska Pipeline System*, August, 1971.

————, *Trans-Alaska Pipeline System Project Agreement*, 1974.

Berry, Mary Clay, *The Alaska Pipeline—The Politics of Oil and Native Land Claims*. Bloomington: Indiana University Press, 1975.

Berton, Pierre, *The Impossible Railway—The Building of the Canadian Pacific*. New York: Alfred A. Knopf, 1972.

Brower, Kenneth, *Earth and the Great Weather—The Brooks Range, Friends of the Earth*. New York: Seabury Press, 1971.

Brown, Tom, *Oil on Ice*. San Francisco: Sierra Club, 1971.

Chasen, Daniel Jack, *Klondike '70—The Alaskan Oil Boom*. New York: Praeger, 1971.

Cicchetti, Charles J., *Alaskan Oil—Alternative Routes and Markets, Resources for the Future*. Baltimore: Johns Hopkins University Press, 1972.

Cooley, Richard A., *Alaska—A Challenge in Conservation*. Madison: University of Wisconsin Press, 1967.

Cooper, Bryan, *Alaska—The Last Frontier*. New York: William Morrow, 1973.

Corrigan, Richard, and Claude E. Barfield, "Alaska Pipeline Lobbying." National Journal Reports, Government Research Corporation, August 11, 1973.

Cotton, H.C., and D.J.B. Thomas, "Super-Diameter Pipes for Alaska." London, England: The Iron and Steel Institute, 1976.

Cotton, H.C., and I.M. Macaulay, "Using Steel in Arctic Construction."
St. Jovite, Quebec, Canada: Conference on Materials Engineering in
the Arctic, September, 1976.

Day, Beth, *Glacier Pilot*. New York: Holt, Rinehart and Winston, 1957.

Fitch, Edwin M., *The Alaska Railroad*. New York: Praeger, 1967.

Garfield, Bryan, *The Thousand-Mile War*. New York: Doubleday, 1969.
Gruening, Ernest, *The State of Alaska*. New York: Random House, 1968.

Helmericks, Harmon, *The Last of the Bush Pilots*. New York: Alfred A.
Knopf, 1969.
Heyerdahl, Thor, *Kon-Tiki*. New York: Rand McNally, 1950.
Human Resources Planning Institute, *A Forecast of Industrial and Occu-
pational Employment in Alaska*. Fairbanks: Institute of Social,
Economic and Government Research, University of Alaska, 1974.
————, *Manpower and Employment Impact of the Trans-Alaska Pipeline*.
Seattle: 1974.

Janson, Lone E., *The Copper Spike*. Anchorage: Alaska Northwest Pub-
lishing, 1975.
Johnson, Paul C., *Alaska*. Kodansha International Ltd., New York:
Harper & Row, 1974.

Kickingbird, Kirke, and Karen Ducheneaux, *One Hundred Million Acres*.
New York: Macmillan, 1973.
Knowles, Ruth Sheldon, *America's Oil Famine*. New York: Coward,
McCann and Geoghegan, 1975.

Larminie, F.G., "The Arctic as an Industrial Environment—Some Aspects of Petroleum Development in Northern Alaska." Journal of Petroleum Technology, January, 1971.
Laycock, George, *Alaska—The Embattled Frontier*, Audubon Society.
Boston: Houghton Mifflin, 1971.

Manning, Harvey, *Cry Crisis! Rehearsal in Alaska*, Friends of the Earth.
San Francisco: 1974.
Maybourn, Ralph, "Sealift to Arctic Alaska, 1975." Great Britain: Polar
Record, 1976.
McPhee, John, *Encounters With the Archdruid*. New York: Farrar, Straus
and Giroux, 1971.

Nathan, Robert R. Associates, Inc., *Implementing the Alaska Native
Claims Settlement Act*. Anchorage: The Alaska Native Foundation,
1972.

Naske, Claus M., *An Interpretative History of Alaskan Statehood*. Anchorage: Alaska Northwest Publishing, 1973.

Nyman, Doublas J., and Thomas L. Anderson, "A Strategy for Seismic Qualification—Trans-Alaska Pipeline System." Draft for future publication, 1976.

Potter, Jean, *The Flying North*. New York: Macmillan, 1947.

Smith, William D., *Northwest Passage*. New York: American Heritage Press, 1970.

Turner, Morris J., "Mobilization of the TAPS Haul Road." Washington, D.C.: Department of the Interior, 1970.

Tussing, Arlon R.; George W. Rogers; and Victor Fischer, *Alaskan Pipeline Report*. Fairbanks: Institute of Social, Economic and Government Research, University of Alaska, 1971.

U.S. Department of the Interior, draft environmental impact statement for the trans-Alaska pipeline, January, 1971.

U.S. Department of the Interior, final environmental impact statement for the trans-Alaska pipeline, June, 1972.

Index

Index   225

Index